11/04

Gearheads

The Turbulent Rise of Robotic Sports

Brad Stone

Simon & Schuster
New York · London · Toronto · Sydney · Singapore

SIMON & SCHUSTER
Rockefeller Center
1230 Avenue of the Americas
New York, NY 10020

DESIGNED BY LAUREN SIMONETTI

For information about special discounts for bulk purchases,
please contact Simon & Schuster Special Sales:
1-800-456-6798 or business@simonandschuster.com

Manufactured in the United States of America

10 9 8 7 6 5 4 3 2 1

Library of Congress Cataloging-in-Publication Data
Stone, Brad.
Gearheads : the turbulent rise of robotic sports / Brad Stone.
p. cm.
1. Robots—Design and construction—History.
2. BattleBots (Television program). I. Title: Gearheads. II. Title.
TJ211.15 .S76 2003
796.15—dc21
2002191244
ISBN 0-7432-2951-7

For my parents

Contents

Above all, the machine has no feelings, it feels no fear and no hope . . . it operates according to the pure logic of probability. For this reason I assert that the robot perceives more accurately than man.

—Max Frisch, Swiss novelist and playwright, 1957

At its base, the dispute is about personal freedom. I am fifty-one years old. Robot Wars was my dream, but it has turned into a nightmare.

—Marc Thorpe, 1999

Prologue

I decided to write a book about robotic sports—and the intensely devoted hobbyists and engineers who participate in them—after watching an amazing championship match at a Battlebots competition on San Francisco's Treasure Island, on Memorial Day, 2001.

I was sitting with 1,500 other spectators in four sets of bleachers that surrounded a transparent 48 by 48-square-foot box. The enclosure was made of a supposedly unbreakable plastic called Lexan. The mechanical athletes were parked inside, waiting.

Son of Whyachi, the rookie, sat in the red square. The 315-pound, remote-controlled robot moved via 16 tiny rectangular feet made of the polymer Delrin, eight on each side, which rotated in an elliptical pattern and shuffled the bot along at three feet per second. Three metal rods extended over the top of the robot and sloped down over the frame, connected to each other by aluminum braces, with steel meat tenderizers at each end. At the start of each match, the apparatus would spin up like a helicopter rotor, whipping currents of air across the floor.

Biohazard, the reigning champ, occupied the blue square on the other side of the arena. The 210-pound heavyweight robot was a marvel of geometric simplicity. Spring-loaded titanium skirts ringed a rectangular frame and extended down to the floor, preventing anything from slipping underneath and gaining leverage. At opportune moments, a stealthy lifting arm would emerge from the base to flip enemies onto their backs or pin them to the wall. It flew across the arena on six wheels at a brisk 15 miles per hour.

Biohazard had speed and dexterity; Son of Whyachi had brute power. Spectators in the stands seemed to know exactly how the match would play out. The champ, they said, would try to reach the rookie and disable it before those destructive helicopter rotors could spin at full speed.

The builders of both mechanical gladiators stood outside the one-inch-thick plastic, nervously shifting their weight from one foot to another, waiting for the match to begin. In their hands, they held radio-controllers, which they would use to send commands to their surrogate athletes over the FM frequency band.

I slid to the edge of my seat along with all the other spectators. Scores of other robot builders were streaming into the hangar from the pits next door, where they had left their toolkits, spare parts, and defeated robots, spread out on row after row of wooden tables.

A tuxedoed ring announcer occupied a lone spotlight at the center of the Battlebox. *"Ladies and Gentlemen, this matchup is for the Battlebots heavyweight competition. Introducing the principals: In the blue square, if you want to take his crown, you'll have to pry it from his cold dead lifting arm. And that ain't going to happen. Your defending heavyweight champion . . . Biohazard!"*

The crowd let out an approving howl. I overheard someone say that in seven years on the robotic combat circuit, Biohazard's maker, Carlo Bertocchini, had won 28 matches and lost only three times. The 38-year-old mechanical engineer from Belmont, California, developed plastic injection molds for a Silicon Valley company, but now that robotic combat was a televised sport—this bout would be broadcast on Comedy Central—Bertocchini was close to quitting his job to build robots full time.

"*In the red square to my right: This robot wanted me to read a letter to his mama. 'If I don't come home with the giant nut, melt me down and give my spare parts to needy robots.' Let's hear it for . . . Son . . . of . . . Whyachi!"*

Broad cheering from the crowd masked a scattering of boos. The guy sitting next to me said that Terry Ewert, captain of Team Whyachi, owned and operated a factory in Dorchester, Wisconsin, that made meat-processing equipment. This was his first robot competition, and he had run his bot-building effort through his company, putting more than $130,000 in parts and man-hours into his creations. "That kind of money will ruin a family sport," the guy in the stands said. Ewert and his team were outfitted in red-and-black NASCAR racing uniforms, which stood in stark juxtaposition to the jeans and custom-made robot T-shirts worn by most of the West Coast competitors.

With the combatants introduced, the announcer hustled out of the arena. The Christmas tree, an electronic display borrowed from drag racing, counted down from red to green lights. A buzzer sounded.

Biohazard lurched forward, but Son of Whyachi hardly moved; apparently its walking assembly had taken a beating over the course of five days and six fights. The helicopter rotor, however, worked just fine. By the time Biohazard reached the center of the floor, the rotor was already a deadly, invisible blur. The two robots collided in a shower of sparks. A square panel of titanium armor disappeared from Biohazard's skirt, landing on the other side of the arena. The robots were flung away from each other by the rotational energy of Son of Whyachi's weapon.

Bertocchini moved Biohazard back into position to take another shot. His best hope was to disable the powerful helicopter blades and then push Son of Whyachi around the arena. The robots collided again. Another small square of armor disappeared from Biohazard's protective skirt, but one of the bracing rods on Terry Ewert's bot also came loose and began to flail around like a piece of clothing sticking out the window of a speeding car. Without the brace, Son of Whyachi's balance was thrown off and the great intimidating helicopter weapon slowed down, then stopped spinning altogether. Bertocchini had an opportunity and we all rose to our feet.

The champ pushed the rookie over to the corner of the arena, underneath one of the four metal hammers known as "pulverizers." These were operated by a Battlebots technician named Pete Lambertson, who sat outside the ring and activated the hazards—the pulverizers, as well as the 16 metal "kill saws" that emerged from the floor—at opportune moments, and *this* was an opportune moment. Lambertson brought the hammer down. It pounded Son of Whyachi's disabled weapon, and the crowd erupted into wild carnivorous cheering, like bloodthirsty Romans at the coliseum. The hammer came down again, then a third time, and a fourth time.

I could see Terry Ewert wincing on the far side of the Battlebox. He was trying to move Son of Whyachi outside the perilous zone, but the walker shuffled ineffectually in place, and Bertocchini's Biohazard blocked all avenues of escape. Lambertson brought the hammer down again, and again. Ewert tried to restart the weapon, but without traction, the base of the robot began spinning instead. We were all cheering and pointing. The bleacher seats shook. The hammer kept banging away. Terry Ewert's face was a vision of pain and disappointment: His $75,000 robot was taking a tremendous beating. Carlo Bertocchini looked serenely confident. A few more hits from the pulverizer and Bertocchini would keep his title and his reputation as the most dominant competitor in the history of the robotic combat circuit.

But Son of Whyachi was still moving, albeit barely. And there was still a minute and a half left in the match.

After what happened next, and all the frenzied scrambling by judges, referees, and event organizers to figure out who had actually won, I started researching all different kinds of robot competitions.

Surprisingly, I found them everywhere. High schools around North America participate each year in the annual FIRST competition, in which hundreds of remote-controlled robots perform specific tasks, such as collecting soccer balls and lifting them into baskets. Computer scientists from Japan gather every December in the Kokugikan sumo hall in Tokyo, pitting their homemade pushbots against each other in fierce one-on-one sumo matches inside a 154-centimeter ring.

At the Georgia Institute of Technology's International Aerial Robotics Competition, students' autonomous helibots fly through the air. At the Trinity College Fire Fighting Home Robot Contest in Philadelphia, machines navigate mock model homes to extinguish a lone candle. In a different city every year, universities from around the world field teams of kickbots in RoboCup, a robotic soccer competition. The ultimate goal of RoboCup, say its founders at the Sony Corporation, is to develop bipedal robots that can defeat the best human teams by the year 2050.

The real competitors in this new sport aren't the robots that roam the field, but the mechanics and computer scientists behind them, who flex their intellects and imaginations instead of their muscles. They're the perfect athletes for an age in which the U.S. Army's drone aircraft fly over foreign soil, robotic rovers explore distant planets, and even cheap toy-store dolls come equipped with hidden computer chips. The robots are among us.

Today, autonomous robot competitions such as robotic sumo and robotic soccer aren't spectator sports. Technologies such as artificial intelligence and computer vision aren't ready for prime time, yet. That's why the most popular and cinematic strains of competition are the more dramatic remote-controlled contests like robotic combat.

At this writing, the concept has been developed into at least a half dozen TV shows including *Robot Wars* on the BBC and TNN and Comedy Central's *Battlebots,* which televised the fight between Son of Whyachi and Biohazard. Some of these shows will disappear. New ones are on the way.

Walking around the crowded pits at Battlebots, I started asking about the history of the sport, but the competitors were hesitant to talk. Sensitive issues involving key figures in the sport were still unresolved, they said. I asked some more questions and learned that the originator of robotic combat was an artist named Marc Thorpe, and that he had been exiled from the world of robot competition.

What I didn't know then—what it would take me months to find out—was that the contests inside the arena mirrored a harsher conflict behind the scenes. That fight took place between people whose proxies were not robots but lawyers, skilled in the martial arts tactics of the U.S. legal system, who extracted a much greater toll than dislodged metal.

My curiosity about that battle resulted in this book, which tells the incredible story of robotic combat, and of the other competitions that are also graduating into spectator sports.

This isn't a book about technology, though it has great technology in it—such as the fearsome machines of the Bay Area's Survival Research Labs, which begin the tale. It's not a book about art either, though many of the robot makers, like the special-effects wizard who created the tiptoeing, crablike robot Mechadon, are inarguably artists, of a type we will be seeing more of in the twenty-first century.

Rather, this is a biography of a sport, the byzantine path it navigated from underground phenomenon to mainstream acceptance, and the price paid for that growth.

If football or baseball were invented today, their trajectories might be similar: Someone has an idea for a new game. The inventor raises money and begins holding matches. A community of enthusiasts forms around the game. The sport appears to have good prospects, and the participants sense the possibility of making money. The players, inventors, and the investors battle for control. They file lawsuits, and the courts have to pick through the rubble. Meanwhile, television takes over and the demands of the small screen (more action, more drama, more carnage) pervert the original idea.

Now that I've finished my research, I think this story is a cautionary tale for *every* inventor or entrepreneur with a new entertainment idea, in an age dominated by television and litigation.

Specifically, two responses to the spectacle propelled robotic combat along its tortuous path. First, almost everyone who saw it, whether under highway overpasses, inside decommissioned military bases, or on television, recognized that it was a great idea, and that it could be really, really big.

Second, everyone who encountered robotic combat interpreted it differently. Some saw it as an art show, some as pure competition, others as "sports entertainment," like professional wrestling. Others concluded it was violent and morally reprehensible, and still others felt it was inherently silly. And no one agreed with anyone else's interpretation.

Perhaps that's why they're still fighting over it.

Part I

The Age of Robot Wars

Chapter 1

The Purist

I'm a parasite. The people I work with are parasites. We live off this putrescent, purulent body of society around us. We scrounge around and we get what we want from them. If they don't give it to us, we take it from them.

—Mark Pauline, 1986

On a cold December night in 2001, a line stretched down an unlit side street in the warehouse district of Berkeley, California. It tailed off at an industrial lot littered with mechanical debris, stretched past a metal fabrication shop and communal art studio called The Crucible, and nosed into a doorway hidden between 40-foot stacks of rust-colored trucking containers. Whatever the impatiently milling crowd was waiting for, it was completely blocked from sight.

There were no mailed invitations or posters for this event, nor any made-up television interviewers, unionized TV cameramen, tuxedoed ring announcers, or really any media attention at all, save for the organizer's own attempts to record the event with digital video cameras. It was all happening below the radar, hush-hush. Only a few days before, e-mails were sent to a reliable few. "Please don't tell anyone," the message urged. "Also, make sure you get there on time, because we're not going to have much time to pull this off."

About 500 Bay Area dwellers showed up. They were artists and hackers, students and computer scientists. Many echoed the same excited sentiment: Finally, after a wait of several years, the robots of SRL—Survival Research Labs—were about to walk the earth once more.

The show was scheduled to begin at 7:30 P.M. In typical SRL manner (and despite the urgency of the e-mail), the guerilla arts group kept the crowd waiting outside for an hour. While the spectators lingered, two uniformed Berkeley police officers walked by, poked their heads into the makeshift doorway, and started asking questions. "What the heck is going on here? What are those . . . things . . . inside?" SRL minions showed their permit, which allowed them to stage a performance art show with a sound restriction of 50 decibels—the level of a normal conversation at a restaurant. "We've done this before. It's an art show. There's no worry here at all, officer," they reassured the police. They wouldn't let the police inside, nor did they produce their ringleader, whose name was explicitly *not* on the permit and who remained safely out of sight. He might be too familiar to the officers. The cops walked away, mercifully unaware of what was planned. And everyone in line who knew about SRL's decade-long struggle with the Bay Area authorities breathed easier.

At 8:30, the doors opened. Spectators paid $20 and were handed a small plastic packet containing two Styrofoam earplugs. Then they were directed through a narrow doorway and into an open yard, where they caught their first glimpse of the spectacle to come. The yard, enclosed on all sides by 20-foot-high stacks of trucking containers, was smaller than the usual SRL show space. The massive, beastly machines seemed nearly on top of each other.

On one side the stationary contraptions were lined up in an orderly row that belied the mayhem they were about to produce. Most conspicuous were the two racing tires of the Pitching Machine. Powered by a 200-horsepower V-8 engine, the wheels would spin in opposite directions, taking two-by-four wooden planks from a conveyor belt and spitting them out at 120 miles per hour, like rounds of machine-gun fire. Next to the Pitching Machine was the long, phallic shaft of the Shockwave Cannon. This fiendish contraption would harness explosions of acetylene and oxygen, directing an eardrum-splitting blast of air and noise that could break windows 700 feet away. Next to that was the mechanism simply dubbed "Boeing." It was an actual jet engine with fuel injectors and an ignition system, designed to produce a long spear of fire—a flamethrower on anabolic steroids. Finally, closest to the audience, was the Pulsejet, calculated to generate a constant thunderous roar of 140 brain-crunching, permit-flouting decibels.

On the other side of the yard were the mobile machines, the lumbering hydraulic robots with names like Track Robot, Running Machine, Inchworm, and Hovercraft. They were all huge, weighing between 500 pounds and three tons. They made Son of Whyachi, Biohazard, and the rest of the Battlebots look like feeble Lego pieces. When the show began, these radio-controlled beasts would wander the yard, menacing the crowd, sparring with each other, and tearing apart the inexplicable SRL props—like the writhing mechanical sculpture called the "Sneaky Soldier," which sat atop a mock lifeguard tower in front of a pool of flammable calcium hydride.

The centerpiece of the show sat at the very back of the yard. It was an electro-luminescent "Geoffrey the Giraffe," the smiling, long-eyelashed icon of Toys 'R Us. The national chain had recently closed a store in the South of Market district of San Francisco and locked two of its Geoffrey signs in a nearby open-air lot. SRL members monitored the lot for weeks, then made their move in the dead of night. This was how the guerilla arts group festooned its shows.

Arranged around Geoffrey's base were large photographs of decapitated heads and jarringly young couples in explicit sexual embrace. Behind Geoffrey, there were photographs of scenic American landscapes. "We had a whole series of effects that were designed to create confusion and extreme sensory overload," explained Mark Pauline, the 48-year-old founder, director, and mastermind of SRL. "Geoffrey was kind of the master of ceremonies."

Pauline wore gray coveralls, the kind psychotic killers always don in slasher flicks. His hair was close-shaven, except, oddly, for his bangs, which were long and brushed back, as if the barber had rushed away on an emergency before finishing the haircut. He wore black, horn-rimmed glasses, and like a general preparing for battle, he walked the yard, calmly giving orders to the 50 or so men and women helping him operate the robots and produce the show. As it began, he donned a pair of industrial-strength earphones and picked up his radio controller with his left hand. His right hand, which he used to manipulate one of the joysticks, had three strangely misshapen fingers and a sac of bulging discolored flesh.

The crowd hurriedly inserted their earplugs as the machines were activated and the mayhem began. The Shockwave Cannon blasted props off the trucking container walls as Boeing torched the photographs of dismembered heads. The Running Machine lurched into the lifeguard tower, and the Sneaky Soldier toppled into the pool of fire. The rest of the robots, creaking with hydraulic motion, circled the yard, clawing each other, shaking their pincers at the crowd. The Hovercraft, difficult to control on the yard's slight incline, boxed itself into the far corner, its four jet engines contributing to the appalling din.

And then the Pulsejet began to scream. Audience members pancaked their hands over their ears, trying to evict the painful ruckus from their heads.

On top of the trucking containers that ringed the yard, SRL's security team monitored police radio frequencies. From Oakland to Richmond, hundreds of East Bay residents were calling 911, reporting what sounded like a ferocious SWAT team firefight near the waterfront. The SRLers heard an emergency dispatcher order a dozen police cars to the scene.

Down in the yard, poor Geoffrey was attacked from all directions. The Inchworm speared him, then Boeing bathed him in flame until the glass plating shattered and that joyful smile disappeared into the burning maw. Then the Pitching Machine was awakened, and in a ratta-tat-tat of wooden planks, eviscerated what was left of Geoffrey's orange and brown spots, until a hydraulic line jammed and the Pitching Machine gagged on its own ordnance. Then the Running Machine tromped over to the flaming mascot and stamped on its remains. Spectators swayed and giggled and clawed at their ears; they didn't know whether to cheer the carnage, like a Battlebots crowd, or to nod knowingly, as if at sculpture in a museum. Or, perhaps, to run for their lives.

Above the yard, the SRL security team heard the emergency dispatcher discover that an authorized art show was in progress near the waterfront. The dispatcher called off 11 of the police cars. Minutes later, SRL security members saw the twelfth car pull up and a lone officer get out. He searched helplessly along the trucking containers for access to the yard but couldn't find the entrance. "How do I get in?" he yelled. From atop the trucking containers, the SRLers shrugged their shoulders and pretended they couldn't hear.

For the show's final flourish, the Pulsejet was turned on the audience. Those closest to the device tried to hide behind each other. Upon discovering that this provided no refuge from the awful din, they stood there and gamely took it, their hair blown back by thrusts of wind. And then, suddenly, it was over. A peaceful calm settled over the charred yard as the spectators looked at each other in wonder and confusion. What the heck had they just witnessed? There was a scattering of tortured applause.

The cop finally found his way in, but he only reprimanded SRL members for starting the show one hour after their permit allowed. Amazingly, Pauline would get no complaints from the Berkeley authorities about violating the decibel-level restrictions on his permit.

As the crowd streamed out of the wasted yard, clutching ears and nursing headaches, the SRL crew started to clean up. Pauline explained the motif of the show, which took place during the second month of the U.S. military action in Afghanistan after the September 11, 2001, terrorist attacks. "We were hoping that people would get the feeling of what it's like to be in a war," he said. "It seemed like the most relevant kind of experience to give people—sort of a Cliff's Notes version of the current state of affairs."

Best of all for Pauline, he had outsmarted his enemies in San Francisco law enforcement across the Bay, who had strained to limit SRL's activities for more than a decade. "I have no bad feelings for the police and fire department," he said. "I'll always find ways to get around the restrictions, and they'll always find new ways to circumvent the shows, and I'll always find new ways to uncircumvent them."

After a while, Pauline wandered over to speak to his mother and her boyfriend. After 20 years of his performances, they had finally made it from Florida to watch a show. Meanwhile, an SRL production manager from Chicago named Deb Pastor climbed up to the top of the trucking containers, where the security team had monitored their police scanners. She hadn't been up there yet and didn't know, as the other SRL members did, that a plastic rain shelter extending from one of the containers could not support any weight. She started to hug a friend, another SRLer named Amy Miller, and took a wrong step onto the parapet.

It crumbled. Both women fell 20 feet, tumbling across Pauline's peripheral vision as he spoke to his mom. "Uh, oh," the SRL chief thought. "That looked bad."

Deb Pastor fell onto a pile of wood. Fortunately, the Chicago production manager stood up and walked away, bruised and scraped but unhurt.

Amy Miller landed directly on the tough earth. The doctors that Pauline always brought to his shows, just in case, rushed over to help her. Other SRLers called 911. A few minutes later, the ambulance arrived. SRL crewmembers rushed the EMS personnel directly through the hidden doorway. A crowd gathered around Miller, who was lying on the ground, moaning in pain.

X-rays would reveal she had broken her pelvic bone in three places and cracked her hip socket. She would need months of rehab and thousands of dollars in medical care. It was not certain that Pauline's insurance would cover it. In the meantime, EMS put her in a hard neck collar and onto a backboard, and shoved her into the ambulance. Then they rushed her to the nearest available hospital equipped to handle traumatic falls, the Eden Medical Center, a half hour away in Castro Valley.

ꝺ

Mark Pauline liked to cause trouble, to transgress the boundaries of convention and safety, and to immerse himself in the world of mechanical things. Those compatible instincts had fueled Survival Research Labs for more than two decades.

Pauline was born in Sarasota, Florida. As a kid, he loved to tinker and take things apart, to test the laws of physics and the limits of machinery. Pauline's parents didn't have a lot of money, and the mean streets of his youth inspired his inventions. He spent a year building variations on the homemade pistol known as a zip gun, experimenting with all the different permutations of materials, from motorcycle spikes to metal cylinders. As a teenager, he built, repaired, and destroyed motorcycles, boats, and cars, and built grenades just to throw them into lakes so he could collect the fish that floated to the surface. Later, he worked for a military contractor building robotic targets, and became an expert welder on the oil fields of Santa Barbara. By the age of 18, he was completely fluent in the art of mechanics.

The love of creation and destruction followed him to Eckerd College in Florida, where he studied literature and experimental theater, and joined the anti–Vietnam War art fringe. "We were all trying to be punked out back in the early seventies," he recalled. At graduation, he cut holes in his red honors gown and underneath it wore a girl's g-string bikini and a pair of brown and white cowboy boots. For final effect, he mixed fluorescent pink paint and Afro sheen in his hair. The professors were outraged, but a tolerant dean let him walk the rostrum anyway.

Pauline arrived in San Francisco in 1978 and spent a few months defacing billboards and pulling other illegal stunts. But he wanted to use his mechanical talents to make his statement, to be an artist while avoiding the staid gallery world. Mostly, though, he wanted to annoy and confuse people—to play pranks and generate real, raw reactions from his audience. "The billboards were like a snack, but I knew there was a bigger meal out there somewhere," he told the Bay Area underground publication *Re/Search* in 1983. "I wondered . . . how can I develop a system that takes all this into account so I can really make a serious assault [on society] and do something that's more satisfying?" Over the course of several weeks, he came up with the idea of staging a mechanized performance. He would take the art of kinetic mechanical sculpture, innovated by artists like Switzerland's Jean Tinguely, and combine it with elements of technology and fire to put the audience in perceived danger. He started sneaking into industrial warehouses and stealing the tools he would need, materials and machinery that would be his clay and paint. For Pauline, the technologies of postindustrial society were its most potent artistic supplies.

He staged his first show, called Machine Sex, in February 1979, at a gas station in the San Francisco neighborhood of North Beach. He didn't ask for permission—he just hopped the fence, plugged his new contraption into the station's power supply, and began. That first apparatus, The Disintegrator, was a conveyer belt that fed objects into a chamber where they got chopped up by a blade, then injected into a hopper that fired them onto the audience. For the show, Pauline dispatched pigeons with his slingshot and dressed them in Arab doll costumes to echo the seventies oil crisis. He put the dead pigeons into The Disintegrator.

The owner of the gas station was mystified and disturbed, so Pauline bribed him with a twenty-dollar bill. "He didn't really say anything: He just took the money and looked at me like I was crazy and then stood back in the audience and watched," Pauline told *Re/Search*. "That's how it all started. That's how I decided to get serious and make playful attacks on society into more or less a way of life."

Pauline stole the name Survival Research Labs from an advertisement in the back of *Soldier of Fortune* magazine. The original owners of the name never called to complain. With a series of collaborators, and then a growing cadre of young cyberpunk followers, Pauline staged mechanical shows in San Francisco through the late seventies and eighties. He moved into headquarters in the gritty, abandoned industrial district of San Francisco. The members of SRL called their base "the junkyard," and called themselves "mechaniacs," for "machine maniacs." The junkyard was stuffed with mechanical tools and parts: 10-horsepower lathes, CNC machines, drill presses and, everywhere, welding gear. SRLers snooped around town searching out the discarded rot of industrial San Francisco, scavenging abandoned materials and tools. Occasionally, they broke into places, helping themselves to parts and equipment. Pauline liked to call SRL a "culture of permissiveness and appropriation."

SRL shows in the eighties were grotesque, surreal, and disturbing. Pauline wanted to jolt audiences and didn't think the machines themselves were sophisticated enough to do it. So early shows featured dead animals: an embalmed rabbit animated by a robotic skeleton, or a mechanical merry-go-round of dead guinea pigs, each flopping back and forth as the whole thing revolved. Shows were performed in parking lots and under deserted highway overpasses, and would occasionally cause audience members to become physically ill.

But Pauline knew that if people were really going to pay attention to what SRL was doing, then the machines would have to be undeniably impressive. Instead of stationary, they would have to move; instead of visibly tethered, they should be radio-controlled. Instead of rusty, metallic, and cold, they should spit fire and blast loud noises. Pauline wanted to really scare, not just sicken, his audience. "If it's technically challenging, I knew they would accept it," he said.

He started experimenting with new technologies, real military and industrial hardware that he bought secondhand from surplus dealers and scrap yards, and in June 1982 he set to work constructing a real rocket motor. The propellant consisted of an oxidizer mixed with saturated fuel, designed to make it burn even faster. He was reading from an instruction manual and mixing the cured solution in a tube when he started having trouble: The manual instructed him to remove the metal rod in the center of the solution, but the rod wouldn't budge, so he went outside his shop and started tapping on it with a hammer. It still wasn't moving. He hit it harder, and it moved a little. He hit even harder, and suddenly he was on the other side of the lot, lying on his back.

"I looked up and I was lying on the ground and blood just went in a sheet of red over my eyes," he said. "And then I shook it out of my eyes and looked at my hand, because it felt funny, and all I could see was just the bones . . . there was no skin on any of the bones. Then they took me to the hospital and put me out and that was *it*."

Pauline was lucky—most of the shrapnel from the fuel explosion missed his head and body. But four fingers on his right hand were gone. Only his middle finger—the universal symbol of vulgar defiance—remained.

It took eight months for doctors to surgically reconstruct the hand, installing two of his toes as replacement fingers. "He's proud of the hand," said a friend. "It's his robot part."

Pauline talked to *The New York Times* about the accident in 1998. "When I examined the situation, I realized I was just another white male who had lived a life of privilege," he said. "Nothing bad had ever happened to me, and I'd gotten a sense of hubris. It becomes destructive when you think you can do anything and get away with it."

When he restarted SRL, Pauline applied his mechanical talents to the task of adjusting. He retrofitted his motorcycle with a second foot-pedal brake so he could stop quickly. He learned to pinch and grab with his surgically altered hand. And he got back to work, with a newfound respect for safety.

Pauline finally did make the rocket, and the new machines were as spectacular as he planned. There was the first Shockwave Cannon, which could break windows hundreds of feet away.

The Air Launcher, based on technology developed for avalanche control, used CO_2 to fire concrete-filled canisters 500 feet per second. During shows, canisters were aimed at prop-houses or other robots, and upon impact, would explode into thick clouds of concrete dust.

There was the Running Machine, a six-legged, gas-powered contraption with a motorcycle-chain drive inspired by a Gatling gun loading system. Pauline salvaged the motors from a defunct brewery, and over the course of several years, painstakingly fashioned all the sprockets himself, an "interesting form of self-punishment," he said. As for the metal that made up the robot's legs and frame, he recalled, "Someone was storing it. I don't remember what happened, but suddenly I had all this aluminum." When it walked, the Running Machine arm articulated a knife back and forth.

"I love making these things," Pauline said. "It's a horrible, painful process, full of all kinds of uncertainty. These are things no one has ever tried to do before. It takes a couple of years to build each machine. By the time you're done, it changes you." Pauline and his compatriots were true hard-core gearheads: They loved the oily complexities of machines and understood the internal mechanics that are, to the rest of us, an inscrutable mesh of wires and servos. While others only dreamed about taking the technology of their daily lives and subverting it into wild, destructive machines, Pauline actually did it.

When he wasn't working on his monstrosities, Pauline bought and resold surplus machinery and computers, earning enough money to support his art, and occasionally send his troupe of mechanical and human performers abroad. In 1988, Pauline took his first major trip, first to New York City and then to Europe. Forty volunteers and collaborators followed him, all of them engineers, artists, and welders who shared Pauline's appetite for artistic destruction—and playing pranks. That May, in front of an audience of thousands in the parking lot outside Shea Stadium, two dozen SRL machines trucked over from the West Coast were set loose in the pouring rain. The show featured the 30-foot-high Wheel of Misfortune, a large spinning circle that rambled over everything in its path, and the one-ton Walking Machine, a lumbering, skeletal elephant that tromped among the debris. During the show, flimsy cardboard boxes were catapulted through the air and burst open over the crowd, scattering three million dollars in counterfeit bills over the parking lot. A friend of Pauline's had found "a Korean who didn't speak much English" who owned a Linotype printer, which manufactured the fake money. Spectators rushed to pick up the bills, which looked impressively authentic.

Pauline, who personally crafted provocative names for each show, called the Shea performance "Misfortunes of Desire: Acted Out in an Imaginary Location Symbolizing Everything Worth Having." The next day, *The New York Times* dismissed it: "The Survival Research team apparently thrives on chaos and on creating, and destroying, nonart. Compared to a 'Mad Max' movie, the effects were nothing special."

The fake money was passed around New York City all summer.

ꝺ

By the end of the decade, Mark Pauline's theatrical displays of fighting robots achieved cult notoriety all over the world. Pauline found the news coverage particularly amusing. "The media can never deny coverage to a good spectacle," he said. "No matter how ridiculous, absurd, insane, or illogical something is, if it achieves a certain identity as a spectacle, the media has to deal with it." Pauline also made videos of his events and sold them at art shows, over the phone, and later on the Internet. The videos were passed between friends, from mechanical obsessive to mechanical obsessive, through the loose network of kinetic artists, mechanical hobbyists, and gearheads.

In California, Pauline's alchemy spawned an entire community of renegade robot artists. Of the hundreds of collaborators and volunteers he trained and supervised, many broke off from the group to pursue their own muses. Matt Heckert, a core SRL member in the mid-eighties, left in '87 to work on computer-controlled sound sculpture, which he called the Mechanical Sound Orchestra. It was a collection of steel, servos, and motors that, when activated all at once, suggested a collaboration of sentient musical beings.

Other SRL members also started groups of their own, with names like Seemen, PeopleHater, and Experimental Interaction Unit. Pauline got a personal kick out of calling these groups "SRL franchises," and he hosted frequent, raucous parties for the community at SRL headquarters. The focal point of these gatherings typically involved blowing up things, such as old television sets.

Meanwhile, outside the Bay Area, Pauline's work was also having an impact. One of his videos fell into the hands of a group calling itself the Denver Mad Scientists Club, an informal band of a half-dozen Colorado mechanical engineers who met through the regular gatherings of the Denver Area Science Fiction Association and its annual sci-fi convention, MileHiCon. While the rest of their comrades sat in rented classrooms and chatted about *Star Wars* and *Star Trek,* the Mad Scientists would stand outside in the hall and talk about their mechanical interests. After a while, they began meeting at each other's homes, working on unusual projects like Tesla coils and alcohol-fueled rockets—the kind of stuff they couldn't do at their staid work environments. John Morse owned a machine shop; Bill Lemieux worked at defense contractor Martin Marietta; Bill Llewellin assembled industrial robots for the Micron Corporation.

Mike Bakula, a computer programmer, was the member who first brought SRL videos to the group's attention. They gathered around his TV set to watch Pauline's antics and were captivated. Here was a guy who didn't just dream about the outlandish mechanical contraptions that constantly occupied the gearhead imagination, he actually built them. "Pauline was building really cool stuff, all expensive and of dubious legality. We loved it," Bakula said.

Sequestered in Denver and unable to catch an SRL show, the Mad Scientists started to think about putting on some variation of a Pauline event themselves. Eventually, they combined the SRL influence with two others.

The first was the 2.70 design competition, held across the country at the Massachusetts Institute of Technology. Since the early seventies, MIT's mechanical-engineering department had been holding a contest for design students each year. Students had six weeks to take a bag of random parts (DC motors, welding rods, constant force springs, and so forth) and build a machine that would perform a specific task, such as herding Ping-Pong balls into a trough. Before each round, the machines had to be smaller than one cubic foot, but after the match began they could unfurl tools such as arms and nets to help accomplish the task. The students then faced each other in fiercely competitive, one-on-one matches. The Mad Scientists caught footage of the 2.70 contest on PBS' *Discover: the World of Science,* which had covered the event intermittently since 1981.

The second influence was an annual happening at MileHiCon, called Critter Crawl, a friendly display of exotic homemade and purchased wind-up toys. The "critters" were judged on their creativity, speed, and design, but the Mad Scientists thought that the competition had gotten stale. For true gearheads, wind-up toys weren't much of a design challenge. The Mad Scientists wondered what would happen if they mixed the Critter Crawl with some of the ambition and showmanship behind SRL and the competitive elements of the 2.70 design contest. The result could be a competition of tethered robotic devices, each with its own weapon and drive system: It would be like robotic boxing.

Gathering at a science-fiction association party one night in the mid-eighties, the Mad Scientists sketched out the parameters of such a contest. The robots would have to be small—and safe enough to perform inside a hotel ballroom if they planned to stage it at MileHiCon.

They eventually created rules: These new critters would fall into two weight classes, two pounds and 20 pounds. Like the MIT machines, the critters would begin each match occupying a volume of less than one foot, but after the match started, parts could unfurl into something larger. The playing field would be a ballroom table, about eight feet in diameter, and the critters would be controlled by electronic tether. The event, riffing off the name of its predecessor, was called Critter Crunch.

The first Critter Crunch was staged in 1986. For the next ten years, the competition was primarily limited to the Mad Scientists and their friends and watched by an enthusiastic crowd of attendees on the last day of MileHiCon. The early machines cost about $100 to build and contained only a slim fraction of the technology of an SRL contraption. But for the Mad Scientists, it was a lot of fun. Bill Llewellin built a critter with a butane-powered flamethrower. Bill Lemieux and Mike Bakula designed the first robot with a pneumatic ram, called Big Punch. A 12-gram CO_2 canister, borrowed from a paint pellet gun, powered its spike weapon. Another early critter emitted water-soluble lubricant to ensnare opponents; yet another featured a mousetrap to clamp onto rivals. The competitors won matches mostly by pushing an opponent off the table, and the ground often did more damage than the critters themselves.

The Mad Scientists enjoyed building their garage robots and exulted in the steadily increasing crowds that turned out to watch Critter Crunch every year. But while the number of competitors and the sophistication of the robots would grow throughout the late eighties and early nineties, the event never evolved beyond its science-fiction convention origins. The robots stayed tiny, constrained by the two-pound and 20-pound weight restrictions. Weapons were kept tame, since the crowd of spectators would often huddle around the ballroom table, only a few feet away from the match. And the competition itself only lasted a few hours and stayed within the context of the larger weekend-long convention.

Much later, when the idea of robotic combat was expanded and commercialized in an altogether different way by someone else, interviewers would often ask the Mad Scientists why they didn't try to develop their version of the concept. "We had a horrible lack of interest in doing something like that," John Morse would answer. "We didn't have any aspirations for making it commercial. We were just having fun." The Mad Scientists didn't even formally copyright their own rules and eventually, other science-fiction conventions copied it. The organizers of the annual DragonCon convention in Atlanta, for example, began to put on their own identical event, called Robot Battles, in 1989.

By the early nineties, fueled by SRL's antics and the various robot combat competitions at science-fiction conventions, robotic spectacle was beginning to seep into the cultural bloodstream: *It was in the air.* But no one had taken that final step—to commercialize it, to actually turn remote-controlled robot fighting into a sport.

⌧

Mark Pauline certainly wasn't going to be the one to do it. He saw SRL as solidly outside mainstream culture, siphoning off its machinery and icons and subverting them for his own artistic ends. He insisted that SRL exist beyond the sphere of practicality or easy consumption, far from the hands of the businessmen who might water down or exploit its ethos of anarchy and confusion. Pauline was attacking society from the outside, giving people experiences they couldn't get on TV or at movie theaters.

In the nineties, the entertainment industry came knocking and tested Pauline's resolve. Alice Cooper's manager asked Pauline to build some machines for the hard-core rocker's stage show. Pauline led him on for a few weeks, then pulled out. Later, one of the filmmakers behind the successful *Blair Witch Project* envisioned a film with SRL robots and wrote an e-mail to Pauline: "I felt it might be an opportune time to let you know of my interest in including SRL as part of my research. If this sounds cool to you, let me know. We'll talk about it in more detail."

Pauline wrote back demanding the following terms. "I would have to write and direct any scene of the film that SRL machines are a major part of. I would have final cut on these sequences. I would be paid an amount comparable to that of a major special-effects company for similar work."

"Mark, thanks for your time," the filmmaker wrote back.

"'No' is one of my favorite words," Pauline said. "I like my status as a marginal character and want to keep it that way."

When somebody tried to use SRL images or video footage without asking (or did anything else to anger the SRL chief), Pauline responded even more harshly. He threatened violence. He publicly smeared them on his website. He directed his legion of young hacker minions to inflict high-tech chaos on them, turning their lives into hell until they surrendered. "We don't deal within the legal system," Pauline said. "We use jungle law."

Inevitably, Pauline began to have problems with authorities. For the first 10 years of SRL, he had managed to stay relatively free of trouble by using his remarkable gift for spinning evasive webs of logic.

"Honestly, officer, we never intended to produce an explosion that big. It just got really really out of control . . ."

"I don't know who put the gas there, officer. It should have been water . . ."

"No, no, that can't hurt your ears. It's a real low noise, it's basically a sine wave. You feel it like you feel the bass guitar at a rock concert."

Onetime collaborator Matt Heckert said Pauline made Bay Area police and fire officials look like the Keystone Cops. "You can only do that for so long," he said.

The cops began to catch up in 1989, when SRL staged a show called Illusions of Shameless Abundance, underneath an onramp to the Bay Bridge in San Francisco. It featured big vats of spoiled food, flammable stacks of pianos clustered around the bridge supports, and the fire-breathing, maniacal robots. At the end of the performance, a large trash bag burst and olive-green cylinders scattered over the ground. Each of them displayed the words, "High Explosive—one-half pound of TNT." They had realistic fuses on one end and were filled with plaster. Like the counterfeit bills outside Shea Stadium, spectators picked them up.

The next day, the sun rose over the Bay Area to find the canisters scattered around the region, and authorities feared the worst. A two-mile stretch of San Francisco highway near the ocean was closed for the day while police checked out six containers on the beach. Another batch was found on the Marin County side of the Golden Gate Bridge. Bomb squads crawled over the city. The authorities finally figured it out and all the local papers reported it: Mark Pauline and SRL were playing another prank.

From there, tensions between the SRL and authorities got worse. The fire department asked SRL to tone down its shows. Pauline refused to compromise and couldn't resist poking fire officials in the eye. In August 1991, he tested a V-1 rocket—a replica of the German buzz bomb used to bombard England during World War II, which Pauline had built to the original blueprints. Three hundred people called the city's earthquake hotline. He performed at the opening of the city's new Modern Museum of Art in 1994 and blasted the eardrums of pedestrians blocks away. The museum almost sued him; the fire department gave him a ticket for endangering lives and property, and asked him not to do it again.

Later that year, CBS News broadcast a story on an SRL event that took place on one of the city's piers, administered by the port commission, not the fire department. But the story included footage of a port commission officer—dressed in fire gear—acting helpless and proclaiming the show "out of control." The San Francisco fire department felt humiliated in front of the whole nation and decided to finally crack down on the artist's antics.

In November 1995, Pauline tried another show, called Crime Wave. It was the first of the super-secretive SRL shows—with no posters or invitations, hush-hush, so the fire department couldn't stop it. Phone calls went out less than 24 hours before the performance. Fire officials knew something was happening but didn't know where.

Twenty-five hundred people turned out to watch the show on Beale Street in the South of Market district, right across from one of the city's new upscale housing developments. The centerpiece of the performance was a hydraulically operated, bound and gagged hostage that writhed helplessly amid the carnage. Catapults chucked flaming bowling balls and bales of hay through the air. SRL's familiar cast of robots, including Running Machine, Drunken Master, and One-Night Robot, tromped through the chaos. A "party jail" on the far hill contained two wildly copulating mechanical figures. Around it, four Unabomber puppets (with the face of the FBI's preliminary sketch of the then-at-large terrorist) rocketed into the sky and exploded. "The props from the show were based on images of crime culled from the popular imagination," Pauline said at the time. The show ended after an actual street sweeper, painted black and outfitted with helicopter blade, emerged on the scene and began spitting green smoke at the panicked crowd.

Fire officials showed up just as the show was ending and brought the police and FBI with them.

The authorities sent the crowd home and questioned Pauline and colleague Mike Dingle for the entire night. They eventually charged the pair with unlawful open burning and using explosives without a permit. They spent a day in jail. Pauline ended up with a $500 fine, a community service sentence, and an arson conviction on his record. He was never supposed to perform in San Francisco again.

So SRL brought its show to Phoenix and Austin, Texas. They did those shows and afterward, officials in both cities told them never to come back, under any circumstances. After the Phoenix show, Pauline told the *Arizona Republic,* "SRL is on a public service mission to spread San Francisco's careless attitude toward life, liberty, and the pursuit of intensity across America."

Pauline took his machines to Japan, where he unveiled the wood-chucking Pitching Machine. He ran it close to the crowd, gunning planks of wood 10 feet in front of audience members at 130 miles per hour. "The SRL policy of making ever more dangerous machines seems a fitting response to those who would have SRL perform safer events," Pauline wrote on his website. Afterward, the Japanese government told him never to come back again, under any circumstances.

When Pauline and 15 cohorts got on the airplane in Tokyo to travel back to the States, a flight attendant tapped the SRL chief on the shoulder and said, "The pilot refuses to fly with you." Apparently other passengers had overheard the SRL crew in the waiting area joking about the Aum Shinrikyo terrorist attacks. The artists were led off the plane and later hitched another flight after they exonerated themselves.

"We tow to the laws of physics, not the laws of humans," Pauline liked to say.

By the year 2000, he was playing an elaborate chess match with authorities. He tried one show in Phoenix, and the San Francisco fire department called ahead and got it canceled. He tried another in Kentucky, and that got scotched too. The Berkeley show in 2001 would be preceded by nearly two years of silence from SRL. Nevertheless, Pauline considered it a triumph. "I get my way in the end," he said.

In 1994, an inventor named Marc Thorpe began mulling the possibility of staging robotic fights as a live sporting event. He hadn't seen Critter Crunch or an SRL show—the idea occurred to him while experimenting with a robotic vacuum—but he knew of Mark Pauline, and one of the first letters he sent was to the SRL chief. SRL *was* fighting robots, and Thorpe felt he needed to get Pauline involved. In the letter, Thorpe wondered if Pauline himself might build a robot to compete in this new mechanical competition.

Pauline didn't respond. What Thorpe had in mind was robotic combat on a contained scale. The robots in Thorpe's so-called Robot Wars could weigh up to 100 pounds. But there would be rules. Weight limits. Weapons restrictions. (A horrible, unspeakable compromise.) When Pauline met Thorpe later that year, he told him, "This doesn't allow anything we like to do."

Rebuffed but determined, Thorpe managed to secure SRL's involvement another way. He hired Pauline to construct the very first Robot Wars arena, and to give a speech during the event. Thorpe even convinced Pauline to bring the Running Machine to show off during the intermission.

What happened afterward required Pauline to employ his unique brand of "jungle law."

In the years that followed, Pauline remained dubious of Thorpe's new sport, even as it achieved unexpected notoriety. Unlike SRL and the Mad Scientists behind Critter Crunch, Thorpe was trying to run Robot Wars as a commercial venture. He envisioned a TV show based on the event and, ultimately, toys modeled on the participants' robots. To make it all happen, Thorpe had a business partner—a New York music mogul. It was way too mainstream for anarchic SRL.

"When you go into business like that, you let the genie out of the bottle," Pauline warned. "You're up against real hard-core, hard-nosed, money-making people. When you deal with those kind of people, you've got to fight their fight. Once you enter into that arena with them, that's the real robot wars, dealing with those robots, the human money robots.

You deal with them on their terms."

Chapter 2
Beginnings

Marc Thorpe's passion for this crazy event just made me say, "I gotta go." I'm always a sucker for passion, and he had tons of it. He's wizardly in some ways. He's a mature nerd, a neat guy. He's also very vulnerable. I don't think he was meant to run a mega-business.

—Caleb Chung, inventor of the toy robot Furby

The Idea was born of a quest for creative fulfillment and financial independence under the specter of declining health. Marc Thorpe was only 46 years old when he was diagnosed with Parkinson's disease.

The Bay Area native was an animatronics designer at ILM (Industrial Light and Magic), the special-effects company of filmmaker George Lucas. From 1979 to the late eighties, before computer graphics transformed the film industry, Thorpe toiled in the model shop in the sprawling Lucas headquarters in San Rafael, California, north of San Francisco. He constructed models for the buildings in the futuristic skyline of Cloud City Bespin in *The Empire Strikes Back* and built the tie bomber that scours an asteroid surface searching for the *Millennium Falcon* midway through the movie. Later in his career, Thorpe worked on robotic effects used in the film *Indiana Jones and the Last Crusade.*

It was exhilarating work and Thorpe loved it, especially at the beginning, when ILM had no competition and mandated excellence at any cost. It was the ultimate geek dream job. Thorpe enjoyed it so much he would stay late at night, wiring fiber optics into star destroyers and working on other models.

After 12 years, standing on ILM's concrete floors for 50 hours a week began to take a toll. During long days, Thorpe felt tired from the heavy strain on his flat feet. He moved to a newly formed division of the company, called LucasToys. It was a small group of four inventors within the company charged with coming up with new toy ideas and licensing them to toy companies. From this new perch, Thorpe saw the commercial strength of the Lucas empire, how the movies planted fictional characters in the popular imagination and the company cashed in by licensing the intellectual property to merchandisers.

In 1993, working in the toy division, Thorpe noticed an unusual tremor in his right hand. He went to a doctor and was diagnosed with Parkinson's. The neurological disorder, a degeneration of the brain cells that produce the essential chemical dopamine, affects one million Americans. Medication relieves the symptoms for years, but over the course of the disease's progress, some victims suffer visible tremors and can have trouble walking, swallowing, and speaking. Parkinson's is the biological opposite of Alzheimer's—it destroys the body while leaving the mind intact.

With that troubling diagnosis, Thorpe faced the prospect of an end to his career—perhaps in a decade or sooner. Even worse, when Lucas closed the LucasToys division later that year, Thorpe was out of work.

He told few friends or colleagues about his medical condition. He chalked this up to an "evolutionary fear of being treated differently." Before his condition worsened and became a factor in how people perceived him, he wanted to establish financial security for himself and his family. He had a four-year-old daughter, Megan, a wife, Dennie, and a mortgaged home in Fairfax, a hilly suburban enclave north of San Francisco. He would have to marshal the full arsenal of his artistic, entrepreneurial, and mechanical talents to provide for their future. He would have to work hard to create something lucrative.

If anyone was equipped to succeed under those conditions, it was Marc Thorpe. He had been tinkering and inventing since he was a kid, and brought an obsessive perfectionism to each task.

Born in San Francisco and raised in the East Bay, his father was a lawyer in New York before he moved to California and became a salesman and later an employee of the state social security office. His mother, an abstract painter of city scenes and landscapes, passed down the creative gene. In high school in San Lorenzo, Thorpe played baseball and basketball but "was never quite good enough to suffer delusions of future athletic success," he said. He graduated high school in 1964 and entered the whirlwind of the sixties San Francisco counterculture.

While his peers listened to the Grateful Dead in the Haight, or rallied against the Vietnam War in Golden Gate Park, Thorpe took a more direct approach to the revolution. He wrote a letter to California governor Ronald Reagan, accusing him of being a fascist; Reagan sent a typed response with a handwritten postscript, scolding Thorpe for his youthful intolerance.

Thorpe lived at home with his parents while attending college at California State University at Hayward—and took care of his crocodiles. He first bought Denver, an 18-inch-long South American caiman, at Woolworth's pet department for three dollars. After a month, he bought a slightly smaller caiman named Lady and constructed a six-by-four-foot wood-and-fiberglass pen in his bedroom. But the crocodiles grew rapidly, and mold from the humidity in the room began spreading on the windows. Thorpe moved out of his room and slept on the living-room couch. After a while, Denver got so big he began escaping the tank. So Thorpe brought the pen into the side yard and built a roof and a ramp for his pets, leaving them to frolic there. Growing ever larger, Denver tried to escape the yard, and Thorpe spent a frustrating day prying him off the chicken-wire fence. Later, Denver evaded the fence too, and after a tense search by friends and family, was found bathing in the backyard fishpond.

Every time Thorpe improved the pen, the crocodiles grew larger and tried to escape. Thorpe, the father of robotic combat, would later play out this same evolutionary scenario with remote-controlled robots.

Denver and Lady were six feet long when Thorpe, heartbroken, shipped them off to Florida. He was moving to Davis, California, to get a master's in fine arts at the University of California. "It was one of the hardest things I've ever had to do," he said.

At graduate school, he worked on the wood and metal sculpture that would later attract ILM managers searching for model makers. Thorpe also found himself on the creative frontier of performance art. One of Thorpe's instructors at UC Davis was the noted California artist William T. Wiley. Like other artists around the country, Wiley was agitating against the centuries-old conventions of art and asking the tough questions: What is art for? What are museums for? Does it have to be so boring?

The self-scrutiny led to the emergence of performance art as a new genre in the Bay Area. Instead of simple exhibitions, Wiley and his students staged shows filled with conversation, theatrical routines, and interactions with spectators. While he broke new ground in the Bay Area arts scene, Wiley also implored Thorpe to open up and challenge himself, to put himself and his audience at risk by moving away from those lovely but lonely sculptures.

In the basement of a UC Davis art building one evening, Thorpe complied. The students were putting on a series of performance pieces. When it was Thorpe's turn, he shuffled onto the stage on his knees holding a box of dog biscuits, with his pants around his ankles. A flashlight dangled between his legs on rubber bands, while an old 45 record player played the wistful Santos and Johnny instrumental, "Sleepwalk." Five dogs wandered toward him from all sides of the stage, including Thorpe's red setter, Murphy. Wiley recalled holding his breath. "We thought he was going to get with the dogs or something," he said. Thorpe crawled around the stage feeding them dog biscuits, the flashlight protruding from his legs below his jockey shorts and casting dreamlike shadows onto the walls. "It was incredible, a waking dream," said Wiley.

Over the next decade, Thorpe staged a number of similar performance pieces. At one show with fellow artists at the Berkeley Art Museum in 1972, he writhed and contorted onstage, holding several fishing lines in his teeth. The lines were attached to a throng of fake, fur-covered spiders, each with bells on the ends of their legs. As the electronic music pulsed, Thorpe twisted the lines around himself, drawing the spiders closer and closer until finally he collapsed in a heap, at which point someone covered him with a quilt of newspapers. Thorpe emerged from the quilt in the costume of a big brown grub and inch-wormed his way off the stage. Wiley called it "a brilliant little piece of science fiction."

Art, Thorpe had learned, could be remote paintings of distant landscapes in refined museums, visited only by an elite few. Or the door to art could be jammed open, with artists working collaboratively in front of an animated crowd. Everyone, even the audience, could add something to this creative elixir.

The problem was that few galleries or theaters were prepared to handle the unconventional performances that Thorpe and his friends were staging. Thorpe found a solution in aquatic theme parks, which had theater stages but weren't fully using them. For three years in the early seventies, he worked to get a government grant for an unusual idea: He would teach dolphins to swim in synchronized patterns. "I thought of animal behavior as a material like wood or metal," Thorpe said. "The process of animal training is very much like wood carving, removing everything that you don't want. I wanted to work with animals, dolphins, and create 'behavioral sculpture.'"

The NEA awarded him $3,000 for the project in 1974, and with his girlfriend, Dennie, he headed to Marineland, Florida, a half-hour out of Jacksonville.

It took a year of constant work and thousands of sacrificial fish to train Betty and Eva. He taught his female bottlenose dolphins to circle their tank like torpedoes, roll over balls, and dance in the water by madly rotating their heads clockwise and counterclockwise to the fifties instrumental hit "Wiggle Wobble." He called it the Dolphin Opera Project, until a reporter asked if he was going to dress the marine mammals in opera clothing. Horrified, he changed the name to the Dolphin Performance Project.

On those delightful summer days, park-goers wandered over to the fenced dolphin tank at the back of Marineland and lined up two or three deep. They saw Thorpe frolicking with his actors, often wearing a dolphin costume himself. Thorpe was rail-thin, with a bushy mustache and a thick mane of dark-brown hair. Toward the end of the project, Thorpe and Dennie borrowed camera equipment from the park and filmed the performance.

Upon his return to the Bay Area, Thorpe got his 16-mm film, *Betty and Eva,* aired on Nickelodeon and HBO. But when he showed it to friends and fellow artists, reaction always fell short of his expectations. "It felt like I had found some new frontier and come back to show what I had discovered. People were enthusiastic, but not much came of it," he said. In the end, the Dolphin Performance Project was creatively rewarding but financially fruitless.

Thorpe concluded that he lacked the necessary business instincts to fully promote his art. He might have the creative and entrepreneurial drive, but he needed the tools to turn his projects into successful commercial endeavors. So, after spending a few years back in the Bay Area sculpting, making short films for TV, and taking construction jobs, he joined George Lucas's ILM to work on *The Empire Strikes Back.* He loved it there, until changes within the Lucas empire precipitated his quest for independence.

After the onset of Parkinson's, Thorpe needed something that could quench the entrepreneurial thirst and, more important, support his family in case he became too ill to work. At LucasToys, Thorpe had devised an idea for a line of fighting toy vehicles—cars, tanks, and trucks with various mounted weaponry. He brought the concept to Mattel, but a similar idea the company had financed, called Wheeled Warriors, had failed in the mid-eighties, so execs rejected Thorpe's version.

Thorpe scratched his head and moved on to the next idea, radio-controlled vacuum cleaners. In 1992, Thorpe spent $700 building a radio-controlled tank with four wheels on each side and a flat top, on which he mounted a handheld, battery-powered vacuum. The idea was to turn a chore into something fun, but the tank, it turned out, hardly cleaned at all. Thorpe persevered and explored the idea of larger, autonomous vacuums for office buildings and industrial spaces. He did some research and found that more than a dozen major vacuum companies were pursuing the same idea, so he gave up that too.

At that point, Thorpe had a plate full of creative loose ends, plus an expensive, eight-wheel remote-controlled tank, which now seemed useless.

He didn't know then about the eight-year-old Critter Crunch in Denver and was only vaguely familiar with Mark Pauline's fiery robotic spectacles. But as he looked at that tank and wondered what he could do with it, Thorpe was on the verge of transforming and reanimating those original ideas. He recalled, "As I looked at it, the eight-year-old boy in me envisioned its potential as a dangerous toy with battery-powered tools mounted on it . . . I had a vision of it cutting its way through a wall. That reminded me of my fighting vehicle toy concept, which brought forth the entrepreneur part of me. It was instantly clear that this was how to make the idea work: I could stage events and invite competitors to build their own vehicles to compete in them. Merchandising revenue from licensing would be the principal revenue stream—that is, toys. The business would own the licensing rights to mechanical athletes that others would build to enter the competitions."

This new sport would operate along the lines of LucasFilm Licensing—establishing characters in the popular imagination, then capitalizing on their likeness. Royalties would be divided between the event organizers and the builders. The event itself would be similar in feel and energy level to the performance art Thorpe had put on in collaboration with William Wiley in the seventies. And instead of crocodiles and dolphins, there would be deadly, radio-controlled fighting machines. With this idea, which he excitedly explained to his family and friends, Marc Thorpe had reached back into his past to blend all his different experiences and interests into a synergistic convergence of community, mechanical art, and sport. It would be a festival of construction, destruction, and survival, and it would harness that creative, mechanical intensity—the gearhead instinct—possessed by so many of his friends and colleagues in the special-effects and art worlds.

For a man whose health was declining, it could also be a salvation.

Now Marc Thorpe needed a name for this idea, and fresh out of George Lucas's empire, he arrived at "Robot Wars." The problem, however, was that the machines Thorpe envisioned were not, technically, robots.

Robots, say the purists, must operate autonomously—independent from human control. The machines Thorpe pictured would have little or no onboard processing power. (In fact, sensitive computer wiring would prove a disadvantage in combat.) Builders would manipulate their creations over the AM and FM radio channels, in the same manner that hobbyists operated radio-controlled airplanes, cars, and boats.

But "RC Car Joust" didn't quite have the same ring to it. "It sounded much better to call the vehicles robots," Thorpe said. He also perceived a great, unquenched public thirst to see and interact with real robots in everyday life—a desire fostered by science fiction, but so far unmet by science.

Americans spent a good part of the twentieth century watching androids on TV and movie screens. The Czech playwright Karel Capek coined the word "robot" in his 1921 play *Rossum's Universal Robots,* in which mechanical slaves rebel against their human masters. Science-fiction writers and Hollywood screenwriters took over from there. Isaac Asimov's science-fiction books detailed the three Laws of Robotics: a robot cannot harm humans, it must obey human commands, and it must protect itself. Early films like *Forbidden Planet* and *The Day the Earth Stood Still* introduced robots to the mainstream filmgoer's imagination, and the friendly and personable C3P0 and R2D2 from *Star Wars* brought them firmly into the spotlight. Suddenly, futurists were predicting the advent of an age of productive, mechanical pets, and an end to human household labor.

The result was a gold rush, the kind that Silicon Valley would later experience with the dot-com boom, as venture capitalists invested more than 400 million in 200 new robot ventures in the early eighties. One of the most prominent companies belonged to Nolan Bushnell, the creator of the Atari game system and Chuck E Cheese, the popular eighties video-game arcade and pizza parlors. In 1983, his Sunnyvale, California, based startup, Androbot, introduced a three-foot tall, cream-colored bipedal android, called Topo, which had a conical head and red luminescent eyes. It obeyed joystick commands from the keyboard of an Apple computer up to 90 feet away and could be programmed to serve drinks and, in later versions, talk. Another Androbot product, B.O.B., short for Brains on Board, carried three 8088 Intel processors, which allowed the machine to think and move autonomously. "A new era in technological history opens," Androbot literature proclaimed, "and a dream that has engaged man's imagination for centuries becomes reality."

It didn't work out quite that well. Bushnell faced daunting technological problems. Early microchips provided little processing power, and since laptops hadn't been invented yet, he also had to deal with issues of power supply, reducing electronic noise, and preventing crashes (which, on a friendly robot pal, might prove disturbing to the children). Moreover, Topos cost more than $1,000, prohibitive for most eighties families, and few worked as advertised.

In 1984, the company laid off most of its staff and Topos were sold to hobbyists for a hundred bucks. The other 200 companies in the new and much-hyped market for household robots met with similar ends. It wasn't until the deep-pocketed Sony Corporation introduced the $2,500 AIBO dog in 1999 that the sci-fi dream of home robots was even minimally fulfilled.

Science-fiction expectations, it turned out, had proven unrealistic, leaving an opening for other nonautonomous machines to assume the "robot" mantle. Thorpe, along with Mark Pauline at SRL, jumped into the breach. Mechanical beasts, operated by humans over radio waves, behave as robots are supposed to. They fight. They breathe fire. They walk the earth, echoing the mayhem of science fiction.

During the planning for first Robot Wars in 1992 and '93, Thorpe visited several meetings of home-brew robot hobbyists to talk about his idea. No one offered an objection to calling these tele-operated machines robots. "They just accepted it," Thorpe said. "They were willing to enlarge their sense of what robots are."

Marc Thorpe had an idea called Robot Wars—now what? It took him two years and $7,000 of his own money to nudge the concept closer to reality. He trademarked the name, created a one-page advertisement and bought ad space in *Radio Control Car Action* and the West Coast magazine *Art Week*. He also sat down and wrote the rules for the sport: There would be three events, he decided.

Escort: Competitor's robot would guide a defenseless drone across the arena, while another robot—a house robot, built and operated by the event organizers—tried to stop the drone from crossing the finish line.

Face-off: One competitor's robot versus another, in a five-minute joust to the death.

Melee: At the end of the competition, all the robots would enter the arena in a free-for-all fight to the finish. The last bot moving won.

Thorpe settled on three weight classes, with the lightweight minimum set at 20 pounds and the heavyweight maximum at 100 pounds. He also decided against allowing the use of explosives, flames, or projectiles. If spectators were going to be sitting close to the action, that would be too dangerous.

With these parameters in mind, Thorpe explored potential venues to hold the event and settled on Herbst Pavilion, a former military hangar at Fort Mason, a decommissioned army base on the edge of the Bay in the San Francisco Marina district. From 1992 to the end of 1993, he reserved and canceled the space twice. He didn't have enough money to put down a deposit, and there wasn't enough interest from prospective robot builders. The ads weren't helping; one respondent explained to Thorpe that R/C hobbyists weren't the proper audience for his competition. While they liked to assemble planes and boats from kits, they weren't necessarily adept at designing machines and weapons from scratch.

By the end of 1993, Thorpe's finances were tangled up with the speculative Robot Wars venture. It was starting to seem like an awfully big gamble.

On New Year's Day, 1994, Thorpe lay on the couch with the flu. He thought about Robot Wars and how he could pull it off, about his declining heath and the high stakes, which kept getting higher. Determined to remain productive, he put together a package of information and sent it to a new Bay Area publication with lots of buzz: *Wired.* Founded not one year before by husband-and-wife team Louis Rossetto and Jane Metcalf, the technology magazine venerated anything and everything related to the mushrooming "digital revolution" in the Bay Area.

An editor called Thorpe the next week. His brainchild was perfect for *Wired*. They wanted to send a photographer right away, to take a picture of Thorpe's own combat robot.

Thorpe panicked. He didn't even have a fighting robot! What he had was the eight-wheel tank with a handheld vacuum cleaner on top of it (which didn't even vacuum properly). That was going to look damn silly in the uber-hip magazine of the future. So he went to a hardware store and bought the coolest-looking chainsaw they had and substituted it for the vacuum. He didn't even bolt the chainsaw to the tank—it just sat there for the camera. After the photo shoot, Thorpe returned it to the hardware store.

Wired writer Jef Raskin called Thorpe at home later that month. Raskin was a multitalented scientist and a member of the gearhead fraternity at its most exalted level. His ideas inspired the creation of the Macintosh computer, and he was a prolific magazine and book writer and an avid maker of R/C airplanes.

In the seventies, Raskin was also a visiting professor at the Stanford Artificial Intelligence Lab, where scientists were building the first generation of autonomous robots, which did little by today's standards. One model, Raskin recalled, was essentially a steel table on four wheels with a video camera and an onboard computer for mapping spaces. One day, the robot was deposited in a Stanford parking lot to transmit images to a TV screen in the lab. That afternoon, Raskin walked past the monitor and noticed what appeared to be cars streaking across the screen. With his colleagues, he raced to the lab's pickup truck. The robot had escaped the parking lot and was meandering down Arestredero Road in Palo Alto.

Raskin loved robots and was initially skeptical of Thorpe's idea. These were not robots, he pointed out, they were R/C cars with weapons. They could be *clever,* but certainly not technologically *deep* in the way that the robots at Stanford's AI Lab were. He was even more concerned about the violence inherent in the idea of garage-built robots bashing each other. "I could think of other, more positive goals for robot designs," he said.

Still, Thorpe's enthusiasm and passion for the event impressed Raskin. Anyone could participate, Thorpe told him, and it would get kids interested in engineering. Just think about it, Thorpe implored:

robo-building as a sport, a contest between analytical minds, with the celebration and adulation lavished on athletes instead bestowed on eccentric, unathletic computer geeks and gearheads.

Raskin was eventually converted. "I understood what its visceral appeal could be," he said, and Raskin convinced Thorpe to get this thing called an e-mail address, so he could properly communicate with a potential community of robot builders.

The *Wired* story came out in March 1994, in the magazine's seventh issue. Over a full-page picture of Thorpe leaning over the chainsaw-wielding tank, Raskin wrote about plans for the first Robot Wars at Fort Mason, on August 20 and 21, 1994. Still anxious about its element of violence, Raskin quoted Thorpe defending himself. "I don't feel uncomfortable about destruction. Promoting combat between robots instead of people is a healthy alternative. That it's aggressive, combative, and survival-oriented gives it a kind of energy that professional football has."

The piece told interested builders to get in touch with Thorpe over the Internet, at Thorpe's new email address, RobotWars@aol.com.

In the following days and weeks, hundreds, ultimately thousands of e-mails poured in. "I spent countless hours going through it, trying to answer the mail, trying to sort it," Thorpe said. "I wasn't sleeping. I wasn't doing anything but working on the event." Where the ads had failed, the combination of the *Wired* clip and the e-mail address succeeded. Thorpe had found his audience.

Now the need for financing became crucial. People were actually building their robots and preparing for the August date. Thorpe owed the deposit on the venue at Fort Mason and would need thousands of dollars more to pull this off. He canvassed friends, who passed word along. Deals feel through. Rysher Entertainment, the production company behind the *Robocop* movies, heard about Robot Wars and was interested. Its president, Rob Kenneally, called Thorpe and sounded excited. "So what will this event be like?" he asked.

"I don't know," Thorpe said. "There's never been anything like this."

"Well," Kenneally said, "what are the robots like?"

"I don't know, they're being built," Thorpe answered.

"So you're asking me to buy a pig in a poke?"

Other financial backers passed too. One friend wanted to front Thorpe the money, then took a financial bath on a simulator ride he was trying to sell to amusement parks. The due date on the deposit for Fort Mason approached. Thorpe saw the prospect for disaster and worried the whole thing would have to be canceled, or worse, end up costing him money that he didn't have. Robot Wars began to look as if it would generate exactly the kind of stress and pressure he should be avoiding with his medical condition.

And then Gary Pini saved him.

Pini was an executive at Smile Communications, a division of a larger independent record label called Profile Records. (Pini's sub-label was among the first companies to use emoticons in its name: the logo of the company read Sm:)e.) Profile was well known in the music industry for helping to usher in rap music in the eighties. When the big labels were ignoring hip-hop, Profile's founders, Cory Robbins and Steve Plotnicki, signed Run-DMC and turned the group from Queens into a popular act.

At the time, the Chicago-born Pini was focused on the wave of techno-music migrating from Europe. His colleague, the London-born disc jockey David "DJ DB" Burkeman, spotted the piece on Thorpe in *Wired* and shared it with Pini. The two high-strung, music-loving New York émigrés went wild over the idea of demolition robotics and saw a synergy with the machine-made synth music they were selling. That spring, Pini called Thorpe to offer encouragement.

They spoke on the phone a few times: Thorpe explained his problems and progress, and Pini offered suggestions. After a while, it became clear to Pini that Thorpe might not be able to pull Robot Wars off, so Pini and DB talked it over. "We decided that we couldn't stand that it wasn't going to happen," Pini said. With $10,000 of DB's cash from working Manhattan nightclubs, they hopped a plane and headed to San Francisco. They met Thorpe at the airport and deposited the money in his bank account, just in time to meet the Fort Mason deadline.

Over the next month, Thorpe and his new prospective partner completed their courtship with a flurry of faxes. Pini took the matter to his boss, Steve Plotnicki, now the sole head of Profile, who didn't quite get the appeal but trusted his employees enough to give his approval. The one thing he asked: Codify Profile's new business relationship with an official contract.

So the lawyers went to work and crafted a joint venture partnership agreement between Marc Thorpe and Profile Records. The original contract was eight pages long and set Profile's initial investment at $50,000, in the form of a loan. Thorpe and Profile would each own a 50 percent share of the new venture—the partners would have to agree on all decisions. Eventually, the contract stipulated, the joint venture would be converted into a limited liability corporation.

Marc Thorpe was ecstatic. Pini seemed affable, supportive, and full of energy and enthusiasm for The Idea. Plus, Thorpe could hand over all financial responsibilities to businessmen who knew what they were doing. Despite his time at Lucas Toys, he had few illusions about his business acumen. For example, he knew little about such routine procedures as "due diligence," which connotes a full exploration of a prospective partner's reputation and business practices. He never thought to research Profile, or the background of its founders. Anyway, Gary Pini seemed like a great guy. With the joint venture agreement, Thorpe felt he was freeing himself to concentrate on creative matters, to fulfill his dream of a new robotic sport.

With the financial pressure alleviated, Thorpe started planning the first Robot Wars in earnest, and one detail he had to contend with was his new company's relationship with the robot-building anarchists of SRL. It was 1994, and Mark Pauline's displays of violent tele-robotics were by then an uber-hyped underground phenomenon. When the technology community in the Bay Area thought of fighting robots, it thought of Survival Research Labs. Thorpe, who had never been to an SRL show, decided to try to involve Mark Pauline.

Thorpe wrote Pauline a letter, telling him about the event, explaining the rules, and asking about his possible participation. Would he like to enter a robot in the competition? Thorpe got no response. He learned a little bit more about SRL and realized Pauline was testing technological and legal barriers with robotic art and would never build such a limited robot. But he respected the artist's commitment and purity of cause and thought about how to involve him in some other way. Serendipitously, to prepare Fort Mason for the event, he hired a production designer with SRL connections. His name was Joe Matheny.

Matheny was a mischievous, severe-looking Bay Area producer with a history of staging successful underground guerilla theater. More recently, Matheny had been working with SRL. When Pauline started having problems getting permits for his SRL shows in the early nineties, Matheny agreed to front the group and get the required certifications from the police and fire departments himself.

Thorpe asked Matheny to handle some of the equipment rental and installation for the event. Matheny would be responsible for erecting the bleachers, blackening out the skylights, and setting up video projection and monitors so attendees could watch the action from all around the Herbst Pavilion.

Matheny subcontracted some of the work to Mark Pauline. The SRL chief remembered Marc Thorpe's letter asking him if he would build a robot to compete. "This is an event that doesn't permit anything that SRL does," Pauline had said. Despite his reservations, the SRL chief took the job anyway.

Later that summer, Thorpe visited SRL headquarters and again laid out his vision. Pauline was in a good mood and agreed to give a speech at the intermission of the inaugural Robot Wars, and to bring one of his robots along for a special demonstration. The deal they worked out stipulated that Thorpe would have Matheny pay Pauline $800 for the speech and demonstration before the event, and another $1,500 installment afterward for the construction work.

On that cool San Francisco summer day, with the first Robot Wars event quickly approaching, Thorpe and Pauline also discussed SRL's escalating problems with authorities. The city was refusing to give SRL the proper permits for its fiery robotic spectacles. "We're just trouble," Pauline complained to Thorpe. "No one wants to work with us."

ꝺ

Summer 1994: The media took one look at Marc Thorpe's plans and devoured the whole thing in a single gulp. *"The media can never deny coverage to a good spectacle,"* Pauline had predicted, and again they were proving him right. There was going to be carnage! Flying pieces of metal! Did they need other reasons to send out a camera crew? Newspaper writers and TV commentators could riff off C3PO and *The Terminator.* These stories wrote themselves.

The event also dovetailed nicely with other changes in the Bay Area technology scene. *If a groundswell of geek energy and entrepreneurialism can create this new thing called the Internet,* the thinking in media circles went, *then we better start paying attention to other groundswells too.* Robot combat smelled like a geek triumph. Mechanical obsessives, converging in San Francisco, were putting on a nerd's sporting event.

Marc Thorpe, unemployed a month ago, now found himself doing interviews, one after another, from his home in Fairfax. *The San Francisco Chronicle* wrote two stories while its local rival, *The Examiner,* did one. *The Chicago Tribune* wrote it up. Even the *Sunday Times* in London and *South China Post* weighed in on the impending event on America's wacky West Coast. Thorpe had an impressive stack of newspaper clips before one piece of armor had been dented. Three weeks before the inaugural Robot Wars, *The New York Times* called. They ran a half-page interview in the Sunday paper.

"What spurred you to the notion that the world needed Robot Wars?" the newspaper asked.

"This is intended as a business venture. I'm partners with Smile Communications and we're serious about it being an ongoing business with long-term objectives. But there's also a lofty objective. If the spotlight of public attention shines on this kind of activity, this kind of design and engineering will get a real shot in the arm. It will have the effect of spawning tremendous interest in engineering and mechanical design for people who are generally distanced from that forum."

"What do you compare it to?"

"It's a sporting event. Tennis used to be only remotely accessible, through country clubs, until it became a sponsored event. The same thing with beach volleyball. When sponsors come into the picture, there are more resources and pretty soon you get more volleyball players. So in this sense, the best thing that can happen from this is that schools will start to have more of a balanced budget between engineering and athletic programs. Not because they ought to, but because of the level of interest that pours in."

"Why do you think this will have that kind of wide appeal?"

"Once you add the element of combat and survival, you are into football fan territory, which is a huge audience. You bring them into the marketplace, then you get Budweiser, Miller, Eveready batteries. It's similar to auto racing. What happens when this kid from Iowa, who spent $300 to build a robot, suddenly gets $300,000 to build a robot to compete against MIT?"

Thorpe was already thinking big.

ꝺ

In the summer of 1994, the new robot builders prepared. No one knew what to expect for the simple reason that nothing like this had ever been done before on this scale. In San Francisco and New York, up and down California and as far away as Madison, Wisconsin, and Lubbock, Texas, they picked up *Wired* or read the newspaper coverage, or they got the e-mails from Robotwars@aol.com. The messages described the new sport, its arena features, classifications, and rules. It laid out the principles behind the escort, face-off, and melee events, and the three separate weight classes. It introduced new restrictions: no explosives, flames, corrosive liquids like acid, or untethered projectiles would be allowed. There could be no radio jamming, electronic weaponry like cattle prods, or liquid weapons like glue or water. Secret "hazards" in the arena would provide other obstacles for the robots. When there wasn't a clear winner, the audience would pick one with its applause.

Gearheads across the country began to work on their response to the engineering challenge this posed. Most of them started late, as few of them knew what kind of time and effort would be required for such a project. And no one knew what anyone else was creating, so they were all engineering in the dark. This first generation of mechanical gladiators was therefore something of a rough draft for the emerging world of robotic combat.

Scott LaValley was an 18-year-old teen from Marin County and a self-described "lost soul." He attended community college, lived with his parents, and didn't know what he wanted to do with his life. In his spare time, he meticulously repaired a classic 1931 Model-A Ford.

His dad showed him an article on Robot Wars in the *Marin Independent Journal*. They worked together for two months, building a trapezoid-shaped tank with tiny Grainger motors, a pneumatic ram and three-eighths-inch-thick aluminum armor. They didn't even have metal tools in their shop, so LaValley cut the aluminum with a hand jigsaw meant for wood, then filed it down. He named his robot DooLittle, after a brigadier general in the U.S. Air Corps who served during World War II.

The name choice would prove prescient. LaValley's homemade speed controller crapped out, so in the final hours before the competition, he scoured area hobby shops trying to find a replacement controller that could handle the current at 24 volts. The one he bought, designed for model airplanes, allowed his robot only to move forward.

Patrick Campbell, 22, was finishing a master's degree in mechanical engineering at Stanford when he read about Robot Wars. He had just completed a class that required him to build an autonomous robot in three weeks. It was challenging and sweetly fulfilling, and hooked him on building robots. A classmate forwarded him Marc Thorpe's e-mail, and since his new job didn't start until that August, Campbell signed up. After graduation, he spent a month and a half building the bot. It was a lightweight with six wheels and two onboard micro-controllers, programmed in the computer language C. Two pneumatic cylinders, powered by carbon dioxide, provided the offensive weaponry and were supposed to drive the spikes forward in dangerous jabs. But Campbell's calculations were off; there wasn't enough thermal energy from the surroundings for the CO_2 to convert from liquid to gas, and the spikes could thrust forward once but wouldn't retract. "Even though I took two thermodynamic courses, I overlooked this at the time," said Campbell. "I should have known better." Campbell called his new robot squad Team Minus Zero.

Charles Tilford, 56, was an inventor in Silicon Valley's Portola Valley when he picked up *Wired* and read the article about Robot Wars. At the time, he was trying to start his own business. He wanted to attend the event, but spectator tickets were $30 a person. Bringing his wife and two young boys, Morgan, 16, and Henry, 10, would cost $150. Why not just enter a robot for the $50 entry fee? he reasoned.

Tilford headed to the scrap pile in his garage and found an old washtub made of galvanized steel. He turned it upside down, bolted on two chains and attached two maces. Inside the tub, he secured two lead acid batteries and a tiny gear motor to make the maces rotate and lash out at opponents. Now all he needed was an actual drive train—the mechanism that moves the robot around the arena. Coincidentally, he was having dinner that week with one of the robotics industry's preeminent experts: Nolan Bushnell, creator of the defunct Androbot corporation. Tilford explained the Robot Wars challenge, and Bushnell offered a solution: Use an old Topo, of which there was no shortage. Bushnell gave one of the failed household robots to Tilford, who stripped it down and used its internal motor for the newly christened South Bay Mauler.

Tilford dubbed himself the "Supreme Commander" of his new robot fighting crew. His neighbor, Larry Bradley, and Bushnell were made "generals." His sons Morgan and Henry, who drew a big, deceptively friendly smiley face on the robot, were "colonels." Charles Tilford protested the Vietnam War in his youth and had practically starved himself to get out of the draft. Now he was middle-aged, struggling in his career, and "not feeling very much like a supreme commander at all." In this world, he would be a warrior.

Many others responded to Thorpe's challenge as well: game designers, entrepreneurs, weekend robot hobbyists. But as the summer progressed, many who undertook the challenge found it difficult to complete their creations and dropped out. Thorpe worried he wouldn't have enough competitors to stage an interesting event, so he called up many of his old friends at LucasFilm and in the toy world. Some didn't need a lot of prodding. Here was a chance for them to ply their mechanical talents in pursuit of athletic glory—and to match wits against fellow gearheads. Others, like Caleb Chung, who would later invent the Furby talking doll and sell 40 million units, wanted to compete but were swamped with other projects. With offers of free plane tickets, adventure, and lifelong friendship, Thorpe convinced them to come.

He recruited other talented mechanical artists to help out in the weeks and days before the event. His neighbor, a robot hobbyist named Dean Simone, built an orange cylindrical house robot, which had a cone-shaped head and was armed with an overhead scoop, like the handle on a sand bucket. The scoop rotated around, swatting and flipping robots into the air. Another friend and former ILMer, Gary Platek, built the first-ever arena hazards to Thorpe's specifications. In the middle of the rectangular ring, on each side, there were two "mousetraps"—hinged square nets that could be lowered onto passing robots. Attached to the side of the ring were also six pneumatic flippers that gave the competitors a little nudge as they careened about. Platek also made the pendulum mechanism that attached to the pavilion ceiling, swinging a bowling ball back and forth over the arena. None of the first generation of arena hazards did all that much, but they looked dangerous enough, and they helped give a field of uninteresting cement the appearance of a high-tech pinball machine.

Meanwhile, various vendors came to Fort Mason and went to work preparing the Herbst Pavilion, a warehouse on a dock surrounded on three sides by the Bay. Joe Matheny's workers covered the sky windows with black visquene plastic and brought in professional lights, lending the pavilion the ambience of a late-night rave. Pauline and his two SRL colleagues built the first Robot Wars arena—a pipe railing two feet off the ground, supported by plywood, that enclosed a playing field 35 by 54 feet.

Gary Pini, DJ DB, and Profile publicist Tracey Miller came to San Francisco a week before the event and set up headquarters in the Phoenix Hotel, a Tenderloin district establishment known for sheltering rock-and-roll performers on their visits to San Francisco. The night before the show, they holed up in their rooms and reviewed Thorpe's rules and plans. "We were making it all up as we went along," Pini recalled. "We didn't know what would happen."

Everyone worked full days, longer and harder than they needed to. While Thorpe's original goals for Robot Wars were commercial, it wasn't predicated on the money, at least for now. It was about proving the sanctity of The Idea—pulling it off, putting on a good show, and (in the mind of a nervous Gary Pini) keeping everyone safe.

Smile arranged the music. Pini and DB brought Ingmar "Dr. Walker" Koch and Cem "Jammin Unit" Oral, from the German synth band Air Liquide. The duo was exploring the musical output of machines. They were, in effect, musical gearheads. The synergy was perfect.

At the entrance to the pavilion, Thorpe exhibited a dozen robots from Bay Area artist Clayton Bailey. Since 1976, Bailey had been salvaging home appliances, car parts, and other mechanical detritus from flea markets and scrap metal yards and converting them into stationary robot sculpture. One robot at the event was called Marilyn Monroebot. It stood more than five feet tall, had two aluminum bats for legs and a stainless-steel coffeepot torso. Two desk lamp reflectors provided the huge shiny metal breasts, tipped with blinking rubber nipples. Thorpe's Robot Wars wasn't just going to be a competition, but a full immersion in the subculture of robot artists.

Seventeen robots entered into the competition on that sunny summer day, August 20, 1994. The event sold out. Spectators lined up to buy the $30 tickets and found themselves beset by the TV cameras canvassing the crowd. Reporters and photographers outnumbered the actual builders. "It was such a natural media event," recalled publicist Tracey Miller. "We knew we were onto something." Thorpe and his crew passed out safety goggles to everyone—and earplugs, for the special SRL demonstration after lunch.

The day started with the lightweight face-offs. If Thorpe had one nagging doubt about the value of his idea, it was that the 10- to 40-pound robots would put on a disappointing show. But then the first two lightweights entered the arena, and the crowd howled with anticipation. AndyRoid, made by puppeteer Bob Cooper and ILM worker Scott McNamara, was a howdy-doody puppet with a red helmet that sat on a Big Wheel, which sported a Batman logo between the handlebars. Mysteriously, it dragged a soda can behind it on a string.

AndyRoid fought Satoru Special, by James Strauss, an animation supervisor at ILM. The robot propelled itself around the arena with spectacular, fiery bursts of a model rocket, though it didn't move that fast. For a weapon, it sported a tiny, front-mounted drill.

The robots were completely ineffectual but hilarious to watch. They hardly touched each other and the crowd cheered anyway. AndyRoid's puppet legs actually moved with the pedals on his Big Wheel. And the Satoru Special emitted bursts of sparks as it jetted across the arena, hitting nothing. Then the bowling ball swung across the floor and blasted AndyRoid in the head. The puppet's red helmet flew off! The capacity crowd let out a great carnivorous roar of approval. And Marc Thorpe knew Robot Wars was going to work.

The other face-off matches followed that morning. Will Wright, creator of the popular video game SimCity, introduced his mechanical curiosity, JulieBot. The middleweight was ramp-shaped and elevated on four wheels. Wright hoped to use a sharpened jackhammer bit on its back as a battering ram. But on top—that's where things got weird. Wright had installed the disembodied head of the talking child's doll called "Julie," once sold by the defunct toy company Worlds of Wonder. Grotesquely, the toy head still uttered phrases like "I'm hungry," "I love a party," and "It's dark in here." Wright painted orange and purple war paint on the doll's cheeks and told reporters the head would serve as a diversion for the other robots.

JulieBot went up against a wheeled disc with a spinning lawn mower blade mounted on top, designed by a Fresno high-school robot team. The high school's robot ran out of power and stopped moving—and the crowd still cheered.

Charles Tilford, wearing a camouflage shirt and black leather jacket, guided his South Bay Mauler into battle. It went down fighting to the expensive and expertly crafted Ramfire 2000, built by R/C car enthusiast and film equipment maker Michael Sorenson. Ramfire's pneumatic piston dislodged Mauler's battery. And Scott LaValley's DooLittle, which only went in one direction, forward, started its first match by sprinting immediately to the other side of the arena and ramming into the wall, where it couldn't move anymore. The crowd loved it anyway.

Then the match of the day: Caleb Chung's Beetle, a wooden box powered by a lawn mower motor, with a protruding pincer meant to grab enemies, versus The Master, built by one of Thorpe's friends from ILM, Mark Setrakian. Setrakian's bot had two bulbous spheres for wheels, one of which concealed a loud, smoky gas engine. Between the spheres, a dangerous saw protruded. Setrakian himself embodied the goth, rave vibe that Gary Pini and DB imagined when they first heard of Robot Wars. While in person he was soft-spoken and modest, Setrakian wore all black, and with shoulder-length dark hair, projected a sinister aura. Setrakian tended to comport himself in the manner of a Vulcan with the journalists and onlookers who swarmed the pits. "What have you got there?" a reporter asked him that morning as he walked into Fort Mason with a box of equipment. "It's parts to a robot," Setrakian answered in his deadpan manner.

Setrakian had worked on robots in the films like *Batteries Not Included* and *Batman Forever,* and was more skillful than most of his robot-building brethren. The Master cut up the competition, and in the match against the Beetle, it sank its saw into the Beetle's wood frame and disgorged a shower of sparks and smoke. The crowd let out another delighted roar. Though The Master would lose the first heavyweight face-off championship to Michael Sorenson's tanklike Ramfire 2000, Setrakian had begun what was to be a prolific career in the new sport of fighting robots.

Then it was time for lunch, and after that, Mark Pauline's speech and demonstration.

Pauline's Running Machine dwarfed the smaller combat robots, and for this special occasion, it sported an earsplitting pulsejet. Throughout the morning's festivities, the Running Machine waited at one end of the Herbst Pavilion, cordoned off by red highway safety cones. Thorpe planned to prepare his guests for the SRL show, and especially, to give the kids in the crowd time to insert their earplugs. That pulsejet, he knew, was loud—capable of emitting 150 decibels of deafening noise. After the face-off competition ended, the announcer let the crowd know: SRL would be performing *after* lunch.

But Mark Pauline had been waiting and watching for too long. He fired up the Running Machine right then. RRRRRRRRRRRrrrrrr! The noise sprayed off the walls of the hangar. Spectators thrust their hands to their heads. Parents grabbed their children and stampeded outside. It was the loudest noise anyone had ever heard in their lives. "We were just screwing around," Pauline said later. "I just ran a machine for a few minutes and everyone got mad. It wasn't a big deal to me."

Half the spectators rushed out of the pavilion. The other half gathered around Pauline to watch. The Running Machine ambled about; bizarrely, two metal pincers rotated photographs of the torso and legs of a naked man. Mark Pauline wore his bulbous earphones and a faint smile of amusement.

Thorpe was exhilarated by the spectacle but annoyed at Pauline's disregard for the spectators. He didn't give anyone time to insert their earplugs. Why did he do it? "Because it was inappropriate to do that, and that is what he does," Thorpe said.

After lunch, it was time for Pauline to give his talk to the crowd. The SRL chief climbed up to the stage and rewarded Thorpe by insulting the builders for the frailty of their creations. "If you're tired of these garage robots," he told the crowd, "come to an SRL event and see some *real* robots."

The crowd booed. They booed because most of them were energized and enthusiastic. The morning action—however anemic these rough-draft robots—exhilarated many of the spectators. They could do this. They could enter and win and have the media crawling around their pit tables, asking them how it felt to compete like a real athlete. There was an electric buzz moving around Fort Mason that day: This could be the sport of the future, they whispered, and they would be the first participants. So Pauline's talk was derided and quickly forgotten, except by Marc Thorpe, who finally agreed with Pauline's own judgment—SRL was trouble.

After lunch, each robot attempted the Escort, guiding a white, mouse-shaped bot across the arena while the house robot tried to stop them. The house robot was heavier than most of the entrants and dominated the smaller bots. Thorpe would drop this event altogether in future competitions.

The final melee provided one last charge to the day. For the first and only time in the new sport, all the robots—from lightweight to heavyweight—entered the ring to slug it out until only one was moving. And it was chaos, utter mayhem. Tiny Tim, a huge forklift that wasn't properly forklifting, bashed the smaller lightweight X1 from Wisconsin. Then it knocked off the antlers that Caleb Chung had stuck on his Beetle. Finally, The Master entered the fray and took out Tiny Tim.

The crowd chanted for the hilarious and feeble Spiny Norman, a lightweight breadbox on wheels with nails sticking out of it. Spiny tried to attack the heavyweight champ, Ramfire 2000, many times its size. It looked to the crowd like a little dog, nipping at the heels of a horse, and they chanted its name. *Spiny! Spiny! Spiny!* Spiny got wedged under Ramfire, whose steering servo burned out, filling the arena with smoke. As Ramfire met an ignominious end, Spiny emerged from underneath and crawled away, to the bleacher-shaking appreciation of the crowd.

Then Spiny Norman wandered into the massive house robot and shed its metal shell in the collision. Electronics exposed, Spiny got crushed underneath the giant's wheels.

The South Bay Mauler also attacked the house robot, and its maces were torn off by the bot's flipper. The house robot proceeded to whip Tilford's robot with its own weapon, to the audience's great and vocal amusement. Spinning around in the middle of the arena, the house robot's amplified weapon took out X1, Zomo, and everything else that wandered into its vicinity.

Finally, at the end of 10 draining, voice-depleting minutes, the house robot ran out of battery power, and the audience saw that the South Bay Mauler was the only robot left moving. With unanimous applause, they awarded the grand prize to Charles Tilford and his crew.

Cameramen and reporters surrounded the jubilant engineer from Portola Valley in the pits. The Supreme Commander, who hadn't felt much like a conquering hero only weeks before, was now shouting to the media crowd around him. In the following days and weeks, he would be interviewed by dozens of news outlets from around the world. Like all the new competitors, he would grow to enjoy the sensation very much.

"One of the things I like about this is that it's not virtual reality, it's not software, it obeys the basic laws of physics," Tilford yelled. "That's the world out here in engineering: Adapt or die."

¤

The event didn't make any money. Thorpe and Gary Pini of Profile Records were going to have to license the concept to bring any real cash into the business. From a public-relations perspective, though, Robot Wars '94 was a triumph. Word was spreading through the media and over the Internet. Many attendees who showed up to watch would tell their friends and compete themselves the following year. Video footage would be shown on the Discovery Channel and attract even more competitors.

Thorpe, however, didn't have much time to celebrate the success. Energized by the first event, he plowed even more time and energy into the Robot Wars business. He e-mailed old competitors and recruited new ones. He reserved Fort Mason for another year and prepared to make the event a bigger, better spectacle. He reviewed the rules, deciding to increase the top weight limit to 140 pounds, and to add a fourth, "featherweight" class, since the tiny bots had proven so entertaining. And he planned for the future of his business under the specter of his declining health. "I worked on Robot Wars as if my life literally dependended on it," Thorpe said. "Nobody really knew the extent to which I was driven."

Only two leftover issues from Robot Wars '94 left a sour taste in his mouth.

The first involved Thorpe's production assistant that year, Joe Matheny, and Thorpe's uneasy cousin in this new world of combat robots, Mark Pauline.

It was bad enough that Pauline strained everyone's eardrums, then ridiculed the competitors to the audience. He was even quoted in a newspaper as derisively calling Robot Wars "an SRL franchise."

Things got worse after the event. Before the show, Joe Matheny paid Pauline the $800 for the speech and robot demonstration. Now Matheny owed Pauline and his two colleagues $1,500 for the arena work. But right around that same time, relations between Pauline and

Matheny grew frosty. Matheny felt Pauline owed him money from securing permits for a prior SRL show and claimed that SRL wrecked the venue, and that the fines ended up coming out of his pocket.

At that same SRL show, hundreds of spectators arrived with fake tickets. Before Robot Wars, Pauline didn't know who to blame for this. Sometime after the event, though, he concluded that Joe Matheny was responsible for printing the bogus tickets and pocketing the money.

In any event, storm clouds moved over their partnership right after the inaugural Robot Wars—just as Gary Pini paid Matheny the last chunk of production cash. Matheny, pissed off at SRL, decided not to pay Pauline the rest of his fee. This was a dangerous move: *The man makes three-ton killer robots.*

"I'll tar and feather you if you don't get me that money," Pauline threatened Matheny.

Matheny decided to lie low for a while. He moved out of his office and virtually disappeared from sight.

Mark Pauline called up Thorpe to demand his payment.

Thorpe and Gary Pini discussed the matter on the phone for hours. Should they pay Mark Pauline, even though they had already given his money to Matheny? Thorpe had worked hard to create a good relationship with SRL and everything had gone wrong. Pini told Thorpe to forget it: They had already paid once. Thorpe was worried because the SRL name carried tremendous cachet in the eyes of the San Francisco cybercommunity, particularly at the influential *Wired*. Did they really want to piss this guy off?

Since Pini held the purse strings, he made the final decision: Mark Pauline was not getting another check from Robot Wars.

Pauline was enraged. On the new SRL website, he slammed the mechanical sporting event. "In view of the insensitive way that an event as flush with cash as Robot Wars dealt with SRL's relatively small payment," he wrote, "Mark Pauline recommends that robot builders stay away from future Robot Wars events."

He treated his new enemy, Joe Matheny, much more harshly and subjected him to a taste of the infamous SRL "jungle law." On the website, he offered cash and SRL paraphernalia for any information on the whereabouts of his former partner. Later, when he got his home address, Pauline printed it on his website and urged his followers to harass him. "Officially," he wrote, "SRL does not encourage or discourage attempts to contact, communicate, or irritate Mr. Matheny."

The legion of hoodlums, hackers, and phone-phreakers who worshiped SRL went to work. That fall, the tires on Matheny's Volkswagen Scirocco were punctured, the mirrors ripped off, and a foul-smelling chemical poured on the upholstery. One morning at three o'clock, the front gate to Matheny's apartment building was chained and padlocked. A note told residents to ring Matheny's bell if they needed to get inside.

And Joe Matheny's phone started to ring, and it didn't stop. One caller told him to report to the SFPD homicide department for questioning in connection with a murder. Others affected wacky voices, screamed obscenities, and hurled sexual threats at his girlfriend. One weekend, the phone rang regularly, every two minutes, day and night. The cops traced the calls to a cracked phone system at an unsuspecting salvage yard in the city's China Basin.

For Joe Matheny, it was a nightmare. But he wasn't going to pay Pauline now. So he spent the better part of that fall sitting on his balcony overlooking the street, a nine-millimeter pistol shoved into his belt, watching over his car and apartment building. Mark Pauline never got the rest of his money from the inaugural Robot Wars, and he never spoke to Marc Thorpe again.

The second lingering issue from the 1994 event concerned a video of the action. One of Thorpe's friends, Greg Becker, the head of an independent video production company, put up $40,000 of his own money to record the first Robot Wars and edit it into a final product. After the event, Becker wrote a proposal for marketing the video, and Thorpe brought it to Profile, which declined to allow it to be sold. Gary Pini and Steve Plotnicki said they were talking to someone else about using the video footage, and they didn't like Becker's strategy for marketing the tape. A few months later, Becker sent Profile a letter, saying he was going to sell his video anyway. Pini and Plotnicki, via a letter from Profile's lawyers, informed Becker he couldn't do that.

The exchange resulted in another round of hapless telephone discussions between Thorpe and Gary Pini. Thorpe couldn't understand what the problem was. He pleaded and cajoled to get permission for his friend to sell the tape but couldn't resolve the impasse. "I was totally perplexed," Thorpe said. "They didn't like the plan to sell the tape but weren't coming up with anything better."

Becker ended up losing his investment. It seemed the "human money robots" might be formidable after all.

Chapter 3

Robot Wars Years

All you had to do was go to an early Robot Wars event, and you knew it was going to be really huge. It didn't take any vision to figure that out. This was what the beginning of something looked like. Everyone who was there was saying, "This is perfect."

—Joel Hodgson, creator of the cable TV series, *Mystery Science Theater 3000*

For a small but growing group of robot hobbyists in the United States, the years from 1994 to 1997 were "Robot Wars" years.

Spectators at the inaugural event were captivated. Many went home and started building a robot themselves for the following year. They told their friends and their neighbors, who told *their* friends, who had possibly already heard about the whole thing via the profuse media coverage. Or, they heard about it later that fall, when "Robot Wars" was featured on the Discovery Channel technology program, *Next Step.* And if those who learned about Robot Wars just so happened to carry in their genetic disposition the specific chromosome that connotes an inclination toward machine design, construction, and ultimately, destruction, then the idea of remote-controlled machines duking it out in a confined arena in front of a raucous crowd was like a shining revelation: *This is it. This is what I've been waiting for.* And they started building too. So, word of Marc Thorpe's new sport spread, through the loose, informal network of people with this kind of mechanical fixation, the gearheads.

In 1994, Thorpe had to gently cajole friends and former coworkers into building robots. In 1995, he didn't have to beg anyone. The e-mails and entry forms started coming right after the inaugural event and didn't stop until the following summer. The new competitors opened their date books to the following year, 1995, and they marked August 19–20 off on the calendars. And the subsequent hours they spent in their garages and basements were often stolen ones. They snuck to their shops late at night or early in the morning, and always, over the weekends. They built their robots in the margins of their day, the hours when their nongearhead brethren sat by the TV watching football and eating snack food. Then the year changed over, the marked-off weekend got closer, and they thought to themselves, *jeez, these robots take a lot of work.* And they stole time until it no longer went unnoticed. The gearheads strained to finish their robots under the resentful eye of whoever it was they should have been spending time with instead. Many matrimonial bonds were tested.

That year, though, they were no longer building in a vacuum. There was the TV coverage of 1994, and a general familiarity with the arena, the rules, and how others had approached the engineering challenge the first time. And there was more time to prepare—a whole year. So the prototypes from 1994 were discarded, and a new generation of combat robots was constructed.

"Supreme Commander" Charles Tilford, for example, put the first South Bay Mauler away in his garage. The old galvanized washtub with the Topo drive train had served him well in the melee. Now he improved on the idea of a spinning weapon: He fabricated a roll-formed aluminum plate and built a drive train from scratch with a pair of motors from two Grainger drills. His sons, Henry and Morgan, made the Team Mauler detail stickers. This time, Morgan, age 17, would drive.

Patrick Campbell, the captain of Team Minus Zero, also returned to the drawing board to craft a new robot, the CyberKnight. He used a stainless-steel frame and mounted a horizontal saw on the front of the robot, with a long rubber V-belt connecting the saw to a gas motor.

Mark Setrakian also reimmersed himself in preparations for Robot Wars. The first Master was built in his spare time at Rick Baker's special-effects studio in L.A.. This year, there was little work to do at the shop and Setrakian was technically unemployed until the next project came along; moreover, he had just broken up with a girlfriend. So he hung around at the shop and worked for three full months, rebuilding his fighting robot from scratch. He built a tail that could spike opponents or hoist them in the air. Instead of a gas-powered saw between the bot's bulbous wheels, Setrakian fashioned several possible weapons, each intended to attack different types of opponents. In addition to the saw, there was a medieval-looking axe, and a nylon board attached to chains and balls, to be whipped at enemies like nunchucks, which would be perfect for the melee rounds. He was turning The Master into a kind of Swiss-army robot.

There were new robots too, elegant and expensive machines that fit within the parameters outlined by Thorpe. Tyler Schilling saw a flyer for the first Robot Wars in '94 and drove 70 miles to San Francisco from Davis, California. Schilling was no mere garage gearhead; he actually owned his own company, Schilling Robotics, which built hydraulic mechanical arms for deep-sea salvage subs and NASA space shuttles. Schilling was captivated by the first show at Fort Mason and, like Terry Ewert of Team Whyachi six years later, directed his employees to prepare a champion robot for the 1995 event.

A half-dozen degreed mechanical engineers went to work. They spent several months during their off-hours building the 140-pound heavyweight, Thor. The mechanical beast was sheathed in brown Kevlar armor and had an air-cooled engine driving a hydraulic pump. The hydraulics powered the front-mounted titanium claw and two rear wheels. The complex robot required two R/C controls to function. Tyler Schilling drove the bot, while an employee, Andy Lyons, controlled the deadly hammer. When the hammer beat back and forth, Thor bucked menacingly, "like a piece of horny farm equipment," said *The San Francisco Chronicle.* Lyons recalled, "We didn't expect to lose."

That August, Schilling and his team drove Thor to Fort Mason in San Francisco and confronted a Robot Wars spectacle twice the size of the year before. This year, instead of 18 prototypes, Marc Thorpe had 49 polished combat robots. The setup was similar to 1994: Packed bleachers surrounded a cement arena in a darkened Herbst Pavilion. But this time, the protection around the arena was more robust. One-quarter-inch thick polycarbonate plastic extended four feet from the plywood base; there was nothing above that. Thorpe and Gary Pini didn't imagine the second-year robots would be powerful enough to send debris over the barrier.

Profile Records brought another German techno artist, Can Oral, (aka "Khan"), the brother of Air Liquide's Jammin Unit, to provide the event's live, loud soundtrack.

Mark Pauline and Joe Matheny, of course, were not invited. Thorpe found a new production manager, Donna Weischel, a former administrative assistant at ILM. Weischel, in turn, recruited friends from her local hangout in Sausalito, Smitty's Bar, to staff the event.

There was another sellout audience that year, and many of the faces in the crowd would, soon enough, be familiar to this growing community of robot warriors. Carlo Bertocchini, the future maker of the Biohazard, was there—waiting, planning his assault. He wanted to make sure that when he entered, he won.

Steve Plotnicki, the head of Profile Records, was there, with his two sons, fraternal twins. The boys were wild about the robots and wanted to tour the pits and see the fighting machines up close. Though he was bankrolling the whole affair, Plotnicki hadn't really understood the appeal of this gearhead subculture. Seeing his kids' enthusiasm was beginning to change his mind.

The second annual Robot Wars spanned two days. The face-off competitions were held on Saturday, and right away the professionally crafted Thor appeared invincible. It doomed Patrick Campbell's CyberKnight, pounding it with its deadly titanium hammer until Campbell's gas motor gave out. Then Thor dominated a heavyweight bot from a team of Marc Thorpe's former colleagues at ILM. The model makers had built a three-and-a-half-foot-wide, pyramid-shaped flipper called Merrimac. Each of the four sides of the pyramid popped up, powered by two pneumatic cylinders. It was intricately designed, well-crafted, and didn't work at all. The flaps gave Thor a gentle push, and the 140-pound heavyweight plopped back on its wheels and continued pounding Merrimac with the serrated hammer, winning its second match handily.

That set up a final, climactic showdown between Thor and Mark Setrakian's The Master. But by that time, everybody's attention in the Robot Wars arena was focused on another warrior altogether.

ꝺ

Like many of his fellow competitors, 31-year-old Jamie Hyneman saw the *Next Step* TV show on the Discovery Channel and fell hard for Marc Thorpe's new sport. A special-effects maker at the San Francisco effects shop Colossal Pictures, he built animatronics puppets for the films *Robocop* and *Arachnaphobia.* Once Robot Wars tweaked his irrepressible mechanical curiosity, he started toying with various ideas for building and entering a robot. As a true dyed-in-the-wool gearhead, a fan of construction and demolition, he decided that he didn't merely want to dominate other robots. He wanted to utterly destroy them.

During those first few months of planning, Hyneman would put himself to sleep mulling his objective. He came to envision a single disc, by itself, spinning faster and faster. A fiercely spinning blade could really rip opponents apart with kinetic energy. Could you get a steel disc spinning so fast at the start of the match that you wouldn't even have to put a motor on it? Well, perhaps in theory. What if you attached a motor to it, and the whole apparatus, including the motor and the armor, spun?

Early robots like the South Bay Mauler featured rotating weapons, but those were skimping on the true destructive potential. Hyneman understood that the heavier the spinning mass, and the faster it rotated, the more kinetic energy would be delivered upon contact with opponent. (Expressed mathematically, kinetic energy equals one-half the the moment of inertia times the angular velociy squared.)

Rooting around in his shop, Hyneman found a powerful turboprop engine, which could produce 20,000 revolutions per minute. What did that mean? Attached to that powerful a motor, how fast would a disc four feet in diameter be going at its perimeter? Hyneman did the math: Its circumference was about 12 feet. Multiply that by the 20,000 rotations, and the perimeter would be doing Mach 5.

Hyneman was suspicious about achieving that.

He called up a physicist friend in his home state of Indiana. "What kind of forces am I dealing with?" he asked. His friend told him that at sea level, the air friction would be so great on any flywheel powered by the engine that the steel of the disc would simply "go off like a flashbulb" and fly apart. So Hyneman revised downward. He would have to spin at lower speeds to prevent his bot from self-destructing, and to withstand hits from enemies without losing balance and bouncing around the arena like a racquetball. At 500 rotations per minute, Hyneman decided, his spinning disc could still do a lot of damage and maintain stability.

It cost only about $500 to build, not including all the free parts he could scrounge from his company's shop. Hyneman and his colleagues turned a metal wok upside down, drilled a hole at the top and painted it black. Around the bottom of the dome, where the bulk of the inertial mass would be located, they attached a quarter-inch-thick plate of steel. In two spots along the bottom of the dome, they bolted steel blades, designed to slam the wheels of opponents while the whole apparatus rotated in a high-speed blur. The dome was placed over a five-horsepower lawn-mower engine. The robot would have little ability to move around the arena, but anything that entered its proximate space would be hammered. Essentially, it would be a passive-aggressive robot—minding its own business until it got attacked, then hitting back hard. It took a few months of work over the weekends until the newly christened "Blendo," weighing in at 140 pounds, was complete.

The spinning blades of robots like the South Bay Mauler "might bruise or break your ankle," Hyneman said. But walk into Blendo, "and it would simply coat the room red."

Hyneman and his colleagues brought Blendo into the Herbst Pavilion at Fort Mason, and the other competitors just stared blankly. This new robot looked like a big UFO. What harm could it do? Hyneman and his crew dressed in fatigues, black Colossal Pictures shirts, and black berets, giving the appearance of a well-honed paramilitary unit. They kept to themselves.

The crowd watched curiously as Hyneman brought Blendo out for its first fight, carrying with him a two-by-ten-foot plank of wood. He placed the plank on the ground, two feet from Blendo, and stood on top of it. He leaned way over his robot, as if he were watering a poison ivy plant, and he poked a power drill down into the hole at the top of Blendo's hemispheric dome. With a quick whirr, the drill activated the robot's engine. The great UFO emitted a deep gasoline belch, then started to spin. Quickly, fearing for his own safety, Hyneman grabbed his piece of wood and scampered out of the arena.

Blendo was fighting a heavyweight named Namreko 2000, built by a welder from Huntington Beach named Michael Okerman. Okerman's bot looked like a coffee urn with corrugated aluminum armor; it ran on the motors of a Barbie jeep toy. On one side, it sported a 12-volt chainsaw, and on the other, a pair of two-inch metal spikes, each sticking out eight inches.

Blendo didn't really move across the arena, it just aimlessly drifted in its blurry, gas-powered whirl. Okerman nudged his creation across the arena and attacked. He directed his bot to poke its dual spikes into Blendo's sphere, to give its enemy a gentle experimental prod, and Blendo warped them. It wrapped the spikes around each other as if they were flimsy pieces of licorice. Later, Okerman would try to straighten them out with a sledgehammer.

Namreko drifted a few more times into Blendo's orbit, and pieces of metal scattered about the arena. After a few more hits, Namreko's battery cable snapped, and it stopped moving. The audience gave Blendo its hearty approval and awarded it the resounding victory.

Back at the pit table, Blendo's makers giddily gave each other high-fives. But Gary Pini was nervous. The Profile executive was spending his time during the second Robot Wars in a shell of mortal fear, panicked that someone could get hurt and lawsuits would follow. Blendo's power justified his anxiety. He sent over a safety inspector, who demanded the team strengthen the bolts on Blendo's blades, which had come loose in the fight.

The other builders, meanwhile, had noticed Blendo's deadly power. They checked the tournament card, ran their fingers down the lineup to see who they faced next and, if it wasn't Blendo, breathed a great sigh of relief. They had built their bots to fight and die, but no one wanted to take his creation home in a paper bag.

The honor of facing the fearsome rookie next went to the kid, 19-year-old Scott LaValley, the maker of the monodirectional DooLittle back in 1994.

Inspired by the first Robot Wars, LaValley had enrolled in classes in computer-assisted design at his junior college. His parents were happy. "They were so excited I was into something," LaValley said. He designed his new bot on the computer and had all of the pieces machined at a nearby shop, siphoning money from his trust fund. The new robot was called DooMore—it went forward *and* backward. The bot was designed around a large aluminum ring, 50 inches in diameter, that LaValley bought from a salvage yard. Two trusses extended across the center of the ring, 10 inches apart, each supporting a large go-cart wheel. Between the trusses, a pneumatic cannon lorded over the entire creation, able to fire and retract a deadly spike. As a secondary weapon, LaValley screwed a series of half-inch bolts, machined to a sharp point, onto the perimeter of the ring. They could conceivably tear at enemies when DooMore spun in place.

LaValley saw that his labor of love was up against the dreaded Blendo. "The whole crew looked like they were from SRL," he said. "They kept to themselves and tried to intimidate everyone. Boy, did it work." He took a deep breath and put his new creation in the ring with Blendo. Once more, Jamie Hyneman brought out that strange wooden plank, so he could stand on it and lean over to start his bot without having his feet chopped off. The drone of the gas motor filled the pavilion, again, and Hyneman grabbed his piece of wood and scooted out of the arena. The tension in the crowd mounted, since no one knew quite what to expect from this madly spinning robot. Blendo began drifting around its side of the arena, waiting, just waiting. And the crowd began hooting and cheering, begging for an engagement. There wasn't much LaValley could do, because his bot was clearly the only one in the ring with navigational control, and this crowd . . . wanted . . . action.

It was a decision all the new robot builders would ultimately face: the choice between the likely annihilation of their painstaking handiwork and fulfilling the responsibility of their sudden role as entertainers. The crowd was cheering! They wanted destruction! Full power ahead!

LaValley nudged DooMore across the arena, past the mousetraps in the center and the swinging bowling ball with the logo of Sm:)e Records on it, and finally, with the resignation of a prisoner being led to the execution chamber, he directed DooMore toward Blendo and fired the pneumatic cannon.

DooMore scored a direct hit. The ram punctured Blendo's shell, which was spinning so fast it yanked the spike off DooMore. Weaponless, LaValley directed DooMore to ram its opponent again. But since Blendo's shell was smooth and round, DooMore slid right on top of it, exposing its undercarriage—and its unsecured batteries, which were attached with duct tape to the back of the trusses. It was carnage. Blendo's blades found the batteries and eviscerated them, turning DooMore into a smoking mess. Worse, the sharpened bolts on DooMore's perimeter started flying off the combatants and out of the arena like popcorn kernels. It was Gary Pini's worst nightmare. Shrapnel cleared the six-foot-high plastic barrier and scattered into the audience. One photographer was nicked harmlessly. Another, larger chunk of metal found the wall near Profile publicist Tracey Miller, who was six months pregnant. "I felt the piece fly by and lodge in a wall behind me," she remembered. "It felt like it was four inches from my stomach."

Gary Pini nearly had a nervous breakdown. He couldn't take it anymore. After the match, as LaValley dragged the ruined carcass of DooMore back to his pit table and Jamie Hyneman celebrated, Pini rushed over to Marc Thorpe and demanded that they disqualify Blendo. "We can't do this anymore," Pini told him.

Thorpe felt bad. It was Robot Wars' fault, not Jamie Hyneman's. They simply hadn't prepared for the possibility that these robotic beasts would exponentially surpass the defenses of the arena—that the crocodiles would grow so quickly and escape their pen—and Thorpe found it difficult to disqualify a competitor who was playing by the rules.

Pini was obstinate, and with none of those gearhead compunctions, demanded Blendo be ousted from the event. "No one is telling me how to run this. I'm going to make a decision. People can't just do whatever they want!" Gary Pini approached the pit table of Colossal Pictures, whose crewmembers were merrily slapping each other on the back. He told them that Blendo was the perfect robot, that it was a wonderful robot, that everyone was very impressed, but that he was concerned for the safety of the audience and, well, how would they react if they left the competition, but got the prize money and the title of heavyweight cochampion?

Hyneman gauged correctly that he didn't have much choice in the matter. He was disappointed, but at least he would get the award. So that was it for Blendo, the first true spinner, in 1995. It got a round of applause from the audience and a $1,000 check for being the heavyweight cochampion—a distinction that was largely honorary.

There was nothing left that weekend for Jamie Hyneman to do, except to watch the proceedings, and when interviewed by reporters, to take fun, competitive jabs at his fellow gearheads. "Blendo is low-tech, but we're proud of that," he said. "We invested less money and effort than anyone in our weight class, and we made a more dangerous robot. I think that means we're smarter."

ꟼ

The 1995 heavyweight face-off came down to Thor, the hammerbot designed and built by the engineers at Schilling Robotics, and The Master, the Swiss-army contraption upgraded from the inaugural event by the L.A. special-effects wizard Mark Setrakian. For the climactic match, Setrakian rode The Master into the ring, one foot in back of the other, standing on it like a snowboard with R/C controller in hand. With his long hair, black clothes, and pale countenance, he looked like a goth king astride his metallic dragon of death. Spectators got on their feet and clutched each other in anticipation.

"It was nerve-racking," Setrakian recalled. "You feel like your life is on the line. The robot is you. It's the sense I would imagine an athlete feels going into the field for an important game."

The hydraulic hammerhead Thor dominated the early part of the match. It met The Master at the middle of the ring and, bucking wildly, smashed The Master's saw with its hammer and stopped the blade from rotating. Setrakian tried to use The Master's new tail to lift Thor, but Thor's hammer slammed wildly up and down, and the robot shook itself free from The Master's grip. Thor got in a few more hits, smashing one of The Master's motors and disabling one-half of its drive train. Left to its own devices, Setrakian's contraption would have aimlessly drifted in a circle until its battery died.

It seemed over for The Master. Thor moved in for the kill, but, at the last possible moment, The Master spun in place, and—a lucky shot?—planted its spiked tail in Thor's side. The spike sliced through a weak spot in the aluminum, which Thor's makers had milled down to reduce weight, and pierced a hole in Thor's hydraulic reservoir. Thor's massive hammer stopped pounding, and the robot shuddered. There was a gentle hydraulic groan and then an exhaust of thick smoke billowed out from underneath the brown armor. Smoke! Hydraulic fluid! The crowd went wild, emitting a vigorous roar that filled the arena. Carving a giant arc in the floor with only one wheel working, The Master pushed the smoldering Thor into the wall and left it for dead. It was a thrilling, unexpected victory for Setrakian. Marc Thorpe would hand the heavyweight trophy to his friend and former colleague that weekend with a satisfied grin.

Now all that remained for Setrakian to complete his triumph at Robot Wars '95 was winning the melee—the communal, fight-to-the-death rumble between the eight still-functioning heavyweights—and one middleweight.

⌧

The one middleweight was called "La Machine"—and its existence worried the heck out of Marc Thorpe.

La Machine was no gas-powered hydraulic monster. It didn't spin, buck, or lift. It was a simple motorized ramp—an unsexy plow on wheels. And it was a team effort.

Back in 1994, Mark Setrakian had invited his old high-school friend and band mate Greg Munson to watch the inaugural Robot Wars. Munson brought his girlfriend, and with no seats available in the packed bleachers, they sat on the floor. He left the event jazzed, but hardly energized like the other spectators who went home and immediately started thinking about what they could build for next year. Munson wasn't really a gearhead. He was a musician, a sound engineer, and to the great displeasure of his parents, he had a proclivity for leaving college to help energetic punk rockers start hopeless bands. Munson's musical career had petered out by the first Robot Wars, and he was working at a San Francisco political advertising firm. After that year's elections, he left and started a CD-ROM design business, Impact Media, with his cousin, Gar Moss.

The next summer, Setrakian's Robot Wars teammate and another high-school friend, Peter Abrahamson, called Munson and told him that he was building an entry for the second Robot Wars event. Abrahamson's new robot was essentially an R/C tank with some customized armor and a few weapons mounted on top. "It's easy," he told Munson. "You should build one too."

Munson started thinking about it seriously, and he invited a friend over to drink beer and brainstorm. They came up with the basic far-fetched robot concepts, robots with fearsome propellers and jet engines and powerful pneumatic punches. The problem was, who would build such a thing? Munson and his friend knew the limits of their mechanical skills. That night, still discussing the possibilities, they brought a bag of empty bottles to the recycling receptacle at the bottom of Munson's live-work loft in Oakland. There, outside the building, they encountered Gage Cauchois. Cauchois, 41, was an eccentric lamp maker who worked from his studio and sold stylized, art-deco fixtures to Bay Area lighting stores. Munson had done some work for Cauchois, soldering parts together for a big shipment of fixtures. That night, by the recycling bins, Cauchois overheard Munson talking about Robot Wars. He interrupted. "Hey, I've heard of that. I saw the *Next Step* show on the Discovery Channel."

Bingo, thought Munson. Cauchois was the real thing, a true gearhead. He could convert their far-fetched ideas into reality. "Do you want to build a robot with us?" Munson asked him.

Cauchois agreed in an instant.

And then, Munson thought of his other cousin, Trey Roski. Roski was a helicopter pilot. He liked to brag that he had an uncanny ability to learn the parameters of a machine and to make it move where he wanted it to go. Roski had just moved up from L.A. to join Munson and Gar Moss at their design business in the Bay Area.

The next day at work, Munson asked Roski if he wanted to participate—to pilot this new, nameless robot they were going to build. Munson gave Roski the *Next Step* videotape and Roski said yes, absolutely, count him in.

Team La Machine was starting to take form.

The cousins started brainstorming, and again they came up with all kinds of destructive robots. But when they presented their ideas to

Cauchois, he said, "No, this is what we're going to do: We're going to build a water-ski ramp." Cauchois, who possessed a more practical, analytical mind, had himself imagined the possibilities. A weekend water-skier, he visualized a giant ski-ramp flying down the highway at 40 miles per hour, pitching oncoming cars into the air. It was simple. The robot's weight would be concentrated in great powerful motors that nobody could match. The ramp would race across the arena toward opponents, which would go right over it, tumbling end over end. Or they would get lodged on top of the ramp, allowing them to get shoved around and slammed against the wall. It was incredible that no one had thought of it already. "From my point of view, we were building a machine that would totally disable other machines, that would take them out," Cauchois said. "What is a robot? Well, a robot is just a machine designed to do a task. That was its task."

While gearheads like Jamie Hyneman wanted to completely destroy the opposition, Cauchois knew he could win by simply dominating them.

Munson and Roski raised $600 for their entry. Roski's mom and Paul MacCready, family friend and the maker of the revolutionary, human-powered airplane glider, the Gossamer Condor, both pitched in.

With those funds, and with Cauchois's simple idea, the new robot was built. Cauchois, a compulsive researcher of tools and materials, found the perfect engines—two motors used to start remote-controlled boats. One motor would control each wheel. He built a cooling fan to keep them from overheating and had them rewound with better insulation wire at a specialty shop in Reno. The bot was covered in plates of sheet aluminum, which were bolted to a wooden chassis with hundreds of rivets. Like many of the early competitors, Cauchois didn't yet know about speed controllers, which allow the driver to control the bot by modifying the speed of each wheel. Instead he built relays, meaning Roski, the driver, would have to quickly turn each motor on and off to turn left or right, making driving all the more difficult. The whole contraption was powered by two lead acid batteries, weighed 80 pounds, and looked like a large wedge of metal-colored cheese.

Greg Munson named it La Machine, after a guitar he had built out of old parts in high school. The guitar, in turn, had been named after the old Black and Decker food processor. "Gage really liked that name," Munson said.

Marc Thorpe hated it. While these mechanical jousts were stretching the very definition of the word "robot," fortunately, that wasn't being pointed out—yet. The name "La Machine," he felt, pulled the curtain away from the wizard at the controls. These were just simple machines, Munson's robot seemed to be saying. Moreover, the robot itself looked boring. Nevertheless, Thorpe kept his feelings to himself.

When they brought their middleweight into the pits at the second Robot Wars, Munson recalled, they looked at the other weaponized robots and felt overmatched. "Where's your weapon?" builders asked them. "Do you think you have enough rivets?" quipped a team of engineers from UC Berkeley.

"We brought it in on a shitty hand truck," Munson said. "We felt totally outclassed."

Then La Machine started to clean up in the middleweight category. It raced across the arena to defeat a bot named Dawn Patrol in less than 10 seconds. It defeated a robot named Scorpion, and in the middleweight championship face-off, beat the first-year winner, X1. La Machine beat each robot in the same way, scooping it up and driving it into the walls of the arena. The next day, it also won the middleweight melee.

La Machine dominated its weight class, but Trey Roski wasn't quite content: There was still the heavyweight melee. The Master, Thor, DooMore, and Merrimac were lining up in the arena for the last rumble of the weekend. While Greg Munson was sitting in the audience and Gage Cauchois was roaming the pits, Roski still wanted to compete. He approached Marc Thorpe and asked if his 80-pound middleweight could go up against the 140-pounders in the heavyweight melee.

"You'll get killed in there," Thorpe told him.

"That's okay, we don't care," Roski said.

Thorpe thought about it. "It took me about 30 seconds to figure out that nobody could object to it," he said. As the final match of the weekend was about to start, Thorpe told the announcer to hold off. "We have another contestant!" the announcer called out. Roski fetched his teammates, they grabbed their robot, and the gates were reopened. As Munson and Cauchois brought in their middleweight, the crowd chanted, "La Machine! La Machine! La Machine!"

La Machine cleaned the gears of its larger opponents. It slammed The Master into the wall. It pushed DooMore into the mousetrap hazard. It harassed Thor like a mosquito, finally nudging it onto one of the pneumatic flippers. Again smoke started billowing up from the heavyweight's Kevlar frame. Finally, at the end of five minutes, no other robots were moving. Roski directed La Machine into the center of the bot-strewn arena, and as the packed crowd at Robot Wars 1995 roared, he sent his wedge into a victorious spin.

It took only a few weeks for Marc Thorpe to discover who Trey Roski actually was.

In November 1994, 10 months before his first Robot Wars, 29-year-old Trey Roski flew his refurbished '77 Aerospatiale Gazelle from his home in Los Angeles to the San Francisco Bay Area. He was going to meet his future in-laws for the first time.

He stopped in Santa Rosa to refuel, then parked his helicopter for the night in the quaint tourist town of Calistoga. The next morning was damp, and when Roski transferred his fuel reserve, water must have seeped into the main tank. Later that morning, a Saturday, with his girlfriend's parents and a copilot on board, Roski brought his helicopter 1,000 feet above picturesque Lake Berryessa. From that altitude, they could see off in the distance the golden foothills of the Sierra Nevada mountains, and directly below, the sprawling lakeside resorts of Napa Valley. Then, ominously, the Gazelle's right motor began to rumble.

Roski decided to play it safe. He brought the copter directly downward, looking for a spot to land on the deserted shoreline. He didn't see the two power lines until it was too late. According to the Napa Valley Register the next day, "Witnesses told authorities they saw the chopper fly low, possibly beneath the lines, then disappear in a cloud of dust."

As Roski described it, the carbon magnesium blades of the Gazelle sliced directly through the cables, sending 115,000 volts of raw electricity coursing through the aircraft. Somehow, the fuel tank didn't explode, but the cockpit window shattered and the helicopter pitched forward. As his earphones flew off his head, Roski cut his power and laid the nose down, bringing the helicopter to a perfect landing on its two skids. "It was the best landing I ever did," Roski said. "It was like the hands of God were lowering us down."

While Roski and his passengers walked away shaken but unharmed, the helicopter was wrecked. The tip of the tail rotor was sheared straight off and every rivet on the aircraft was fried. A passing boat took them to the nearest resort. Eighteen thousand local residents lost power that day, and Roski's dreams of running a helicopter passenger business were officially over. It seemed he had no choice but to go back to working for his dad.

Roski's father, Edward Roski, Jr., was the CEO of Majestic Realty. The company, according to market research firm Hoover's, was "the king of the golden state."

Majestic designed and developed office buildings and industrial parks around the country. Ed Roski, Jr., also owned part of the Los Angeles Lakers and Los Angeles Kings sports franchises, and would later build the Staples Center, where both teams now play. *Fortune* magazine regularly placed Ed Roski on its list of the richest Americans, estimating his fortune at around $900 million.

Overweight since he was a child, with a curly thicket of rust-colored hair, Edward "Trey" Roski III grew up the child of wealth and high expectations. His grandfather built the family empire by acquiring rail yards and building office complexes on the old lots. His father served in Vietnam and rarely spoke to his son and two daughters about that period of his life.

Trey Roski was raised in a home near Lake Toluca, outside L.A.. In the second grade, he was tested and found to have severe dyslexia. His parents flew in experts from all over the country to diagnose him, and he was poked, prodded, and tested in every conceivable way. "This type of disability is unseen," Roski said. "I have all my arms and legs. I speak well. But I never read a magazine. I never get any mail." His parents eventually sent him to the Gow School in Buffalo, which promised to prepare dyslexic boys for college. "Finally, I figured out that I wasn't the only one," he said.

Roski graduated from boarding school in 1983 and made it through California State University at Long Beach in seven years, refusing to surrender to his disability, hiring others to read his textbooks to him and getting books-on-tape from the Braille Institute. He had a cumulative 2.0 average, less than the required 2.2, but the school let him graduate anyway. Then he went to work for his dad. He started in construction and moved up to property management, driving to Majestic properties, checking in on tenants, and making sure the building's roofs were clean. In his free time, he thought about helicopters.

Ever since he was a child, Trey Roski had loved controlling mechanical things. He set up obstacle courses in his bedroom for his expensive remote-controlled cars, and on vacation with his two cousins, they raced R/C boats across lakes. After he graduated college, Roski's obsession shifted to flying. He wanted to get a helicopter license and start a business ferrying people short distances. "I knew I needed to do something that didn't incorporate reading," he said.

The problem was that pilot training cost thousands of dollars, and Roski's parents strictly controlled their children's finances. But Roski didn't want to tell his parents about his dream, then humiliate himself by failing the pilot test because of his dyslexia. Without telling anyone, he took out a $7,000 loan on a life insurance policy that his grandfather had bought him. He used the cash to take flying lessons and passed the exam easily. Then he told the news to his father and grandfather. He hoped to shock them with his accomplishment, then ask them to buy him a helicopter.

The plan didn't quite work. His grandfather was furious and didn't speak to him for months. His dad was skeptical. It wasn't until the younger Roski rented a helicopter and flew his father to Majestic Realty's Silverton Hotel & Casino in Las Vegas that his dad agreed to provide the funds to buy the used Gazelle. Roski fixed it up himself, adding a stereo and a fine cherrywood interior. "It was sweet," he recalled.

His budding business showed signs of promise before the accident. Roski ferried around a small-time rap group making a music video and took the mayor of Napa Valley on a land survey. After the accident, the FAA tried to take away Roski's license, but his family's lawyers managed to prove that the PG&E power cables weren't properly marked on aerial maps, and that this had caused other accidents and even some fatalities in the past. Roski escaped with his license, but he didn't have the money to buy another helicopter. His parents certainly weren't going to buy him a replacement.

A few months later, Roski's cousins asked him to move north to San Francisco and join them in their CD-ROM design firm. By then, Roski knew his future in the family business was limited. "So much of what my dad does is reading over contracts, picking over the fine legal points," he said. "I didn't think I would be able to do that." Trey Roski decided he needed a change of scenery.

Eventually, Greg Munson asked him to drive La Machine in this thing called Robot Wars, and Roski reveled in the unexpected, exhilarating triumph over the heavyweight bots in the melee. It was a dramatic victory, and Roski was enthralled by his success.

For thirty years he had lived in the shadow of his famously achieving father and grandfather. Now he had found *his thing,* something *he* was successful at. Even after the event ended and everyone went home, he didn't want to give up the feeling. The same relentless tenacity he applied to overcoming his learning disability and graduating college, he brought to involving himself in this new sport.

He contacted Marc Thorpe and offered to prepare a brochure for the next Robot Wars. He also helped put together a video of the action from '95, and he called Gary Pini to ask how he could be involved in the business side of Robot Wars. One day that fall, on a trip to San Francisco, Pini stopped by the cramped apartment of Gar Moss, from which the cousins were running Impact Media. Pini, a New Yorker and a music man, didn't understand exactly what the enthusiastic Californians had in mind with their offers of "help." "What you want," Pini finally concluded, "is equity."

"Sure," said Roski, who had never heard of "equity" and had no idea what it meant.

"How much do you want to invest in Robot Wars?" Pini asked.

Roski threw out a number. Two hundred thousand? Three hundred thousand? He figured he could get access to his trust fund money for this kind of thing.

"That isn't enough," Pini replied. "Plus, Robot Wars isn't for sale."

Roski wondered out loud, "What's to stop me from doing it on my own?"

Pini laughed. "We have two years of a jump start on you. Go ahead and do it."

Pini did agree, however, to think about allowing Impact Media to build a Robot Wars website on the new fashionable medium called the World Wide Web. He took the matter to his boss, Profile Records chief Steve Plotnicki, who said he would allow the website but wouldn't pay for it. Plotnicki didn't think much of the commercial possibilities of the Internet and felt the whole Robot Wars venture was getting expensive anyway.

Undeterred, the cousins started working on the site. Sometime that fall, they learned that David Letterman might be interested in doing a segment on Robot Wars. Roski called Profile publicist Tracey Miller and begged her, "Get me on the show!"

Miller told Roski, "It's not as easy as just sending in a tape of the competition. They need to know that the robot builders are interesting and fun people."

Roski decided to be proactive. He invited Marc Thorpe to come along, and with Munson wielding a video camera, they took La Machine to the streets of San Francisco. They told pedestrians they wanted to sell the robot and asked for suggestions on how to put it to use. "A portable skateboard jump," someone suggested. So they filmed skateboarders jumping off La Machine. "You could use it at the airport, to wedge underneath airplane tires, so you don't have personnel running everywhere," someone else advised. Roski called South West Airlines at Oakland Airport and harangued them into letting him film on the tarmac. The cousins showed up early one morning and shot La Machine rolling around planes and luggage trucks. And Roski called the Bay Bridge Authority and asked the same thing, could they film on the bridge? The bureaucrats demurred, but Trey Roski *would not take no for an answer,* he was utterly tenacious, an opener of padlocked doors. The authority finally agreed to let them trek out to the Bay Bridge at six o'clock one morning and film the robot passing through the toll plaza. The video closed with La Machine on a leash, led down Market Street by Marc Thorpe in a lumbering gait that suggested the creeping progress of his disease.

It's unclear whether David Letterman or his staff ever watched the video. The segment never happened. But Thorpe and Roski became friends, and Thorpe learned Roski's background and was thrilled. "You won't believe who this guy is!" Thorpe excitedly told Gary Pini over the phone. A competitor with money and connections to the mainstream sports world could only help robotic combat.

¤

The British, in their typical manner, have an altogether different word for gearhead: "anorak." Literally, it's an unfashionable hooded nylon coat. Anoraks were often worn by train spotters, hobbyists who endure all sorts of weather to record serial numbers from the sides of passing locomotives. Over time, the term "anorak" evolved to pejoratively describe any individual with a boring hobby.

Tom Gutteridge, then, was no anorak. He was born in Newcastle, England, and ever since he could remember, wanted to go into the TV business. After attending university at York, he joined the BBC, where he toiled for 12 years in the creative department before leaving to create a documentary about music with legendary Beatles producer George Martin. Within two years, Gutteridge's company, Mentorn Films, was the largest independent production company in the United Kingdom.

In 1994, a friend named Keith Ford showed Gutteridge a videotape of Marc Thorpe's inaugural Robot Wars event in San Francisco. While he professed to be mechanically clueless himself, Gutteridge loved the spectacle, and in March 1995, invited Steve Plotnicki and Gary Pini from Profile Records to London. Over dinner, he bought the right to create a unique Robot Wars TV show for the United Kingdom, modeled after the live event. Mentorn would pay the Robot Wars joint venture $5,000 for each episode they produced, and 50 percent of the net revenues from all resulting licenses. With that prospective windfall, it seemed that the Thorpe–Profile Records partnership had a promising future.

The following January, in 1996, Profile told Thorpe to gather a bunch of robot makers for a trip to London; Mentorn Films wanted to make a Robot Wars TV pilot and present it to the BBC. They wanted Marc Thorpe and the top American robot builders to come to England and help put on the show. Thorpe asked his new friends at team La Machine. He also brought Mark Setrakian, Peter Abrahamson and The Master, and Tyler Schilling and Thor. Meanwhile, one of Gutteridge's employees in the United Kingdom, Steve Carsey, spent several months contacting European engineers and businesses, asking them to build robots for the pilot. He convinced a technician with the BBC named Derek Foxwell to build three robots to round out the performance.

Suddenly, Trey Roski and Marc Thorpe were in England, proselytizing for their joint love, robotic combat. All the British TV networks wanted to interview Thorpe about this crazy new American import. He stood outside a central London warehouse one cold morning and did interview after interview. They asked him all the routine questions, how he thought up the idea, where he thought it was going, and of course, did he think it was too violent?

Since *Wired*'s Jef Raskin had asked him this a year before, Thorpe was getting quite adept at molding permutations of his answer to the ubiquitous, nagging question.

Is Robot Wars violent?

"This is the kind of violence that's like a bat hitting a ball. There is no human suffering involved, no intended human violence."

Is Robot Wars violent?

"Is football or boxing violent? If they are, this is much less so. No one's getting hurt and it's bringing students into the field of engineering."

But is Robot Wars violent?

"Well, it is violent, yeah."

That night, with a sparse collection of press and TV execs gathered around, the competitors put on a tremendous show. They played a game of robotic soccer, handily won by La Machine, which pushed all the discs—and the other robots—into the net. In the face-off competitions, the American robots systematically defeated Derek Foxwell's three Eurobots, pushing them over the edge of an elevated arena floor. Then the Americans went after one another. The Master and Thor fought a vicious rematch of the 1995 heavyweight final. Setrakian was delighted to have another shot at Thor—he felt he escaped the first battle with a lucky hit. He armed The Master with its original weapon, the saw, and closed in on Thor, precisely targeting a weak spot in the armor and slashing a hydraulic power line, sending up a shower of sparks and smoke. The attack left Schilling's robot in slumbering, smoldering ruin. (After that fight, Schilling would put Thor on a shelf at his corporate headquarters in Davis, California, and never fight it again. It simply took up too much of his employees' time, he said.)

The Master advanced to fight La Machine—and got scooped up by the motorized plow and pushed over the edge of the arena floor. La Machine had triumphed again. Roski and Munson were jubilant. They took La Machine out on the town with a video camera, filming themselves and their robot in front of Buckingham Palace and crossing Abbey Road in the famous Beatles album pose. La Machine took Ringo's position.

Tom Gutteridge recalled that after La Machine's victory in the soccer match, TV execs from the BBC actually smiled—an occurrence that, Gutteridge assured everyone, was rare. Nevertheless, it would take the BBC more than two years to give Gutteridge approval to stage and film a Robot Wars TV show in England.

Trey Roski returned to the United States more excited than ever. Robot Wars seemed predestined for greatness. His simple wedge appeared indomitable. He may not have been a naturally predisposed gearhead, but with a passion and tenacity fueled by winning, he was going to participate in a major way. He would do everything he could to forward his prospects. He would use all the connections at his disposal, and he would be unrelenting.

That spring, he called NewTek, the San Antonio, Texas–based maker of the desktop video-editing software that Impact Media used. Roski explained this new sport to its founder, Tim Jenison, and asked for money. He secured $10,000 for Team La Machine, the sport's first major sponsorship.

Two months after the trip to London, Marc Thorpe had another proposition for his hard-core robot warriors: They were going to Germany. The Smile label of Profile Records was experimenting with the synergy between mechanized sport and machine-made music. It signed up with German promotional firm Avantguard and got Marlboro cigarettes to sponsor a rave tour across seven cities in Germany, right at the end of the European winter. Marc Thorpe would direct the robotic aspect of the production, and received enough money to bring two robots and their builders. Thorpe recognized an opportunity to use his performance art background to stage something inventive and interesting. He asked his friends behind the robot La Machine and young Scott LaValley of DooMore to come along.

They met at the Marin Brewing Company north of the Golden Gate Bridge the night before they were due to set out, and Thorpe laid out his plan. The Smile DJs would be playing most of the time, so they would have 20 minutes to fill each night. Since they only had two robots—La Machine and DooMore—and the facilities wouldn't be suitable for real combat, they would have to improvise. Thorpe told his performers that he was going to buy a dozen small R/C cars to use as props. The cars would swarm over La Machine in a mock fight. He was also going to pick up a few small toy electronic robots and allow DooMore to circle the arena area knocking them over with its pneumatic cannon.

He would also buy a big crate of lettuce heads for another scene in the show. He would attach one head of lettuce to each R/C toy car by a length of rope. Then, two of the robot builders, wearing red reflective vests and armed with sledgehammers, would pretend to try to crush the lettuce as the cars zipped by. It was mechanized performance art, to go along with the pounding techno music, and to prepare the crowd for the climactic match between La Machine and DooMore, although the robots wouldn't truly be able to fight, because there wouldn't be a safe arena. The two robots, Thorpe said, could regard each other for a few minutes and take gentle theatrical jabs, sort of like the combatants in *The Nutcracker.*

Roski, Greg Munson, and Scott LaValley regarded all of this with skepticism. It was . . . strange. But heck, they were going to Germany, getting paid while promoting the sport they loved. Gage Cauchois, however, wasn't as enthusiastic. Greg Munson's neighbor and La Machine's builder was more particular, and after drinking a few beers, became antagonistic. He scoffed at Marc Thorpe's ideas. As he drank even more, he started ridiculing Thorpe and interrupting him. He didn't even want to hear the various job descriptions Thorpe was outlining. He didn't care if they were leaving the next night! Then he demanded to know who was going to drive the bus. He once had been in a bad car accident in Germany. Germany was dangerous! And God forbid that Trey Roski would be driving, Cauchois proclaimed. The man liked to tailgate in his SUV at 85 miles per hour. He bragged about a helicopter crash! And he was supposed to be a good driver?

Cauchois was being his ornery gearhead self, but Thorpe couldn't take it. Later that night, back at their homes, he got Munson and Roski on a conference call. There was no way he could handle Cauchois in a cramped bus for two straight weeks in Germany. Munson went downstairs to tell his neighbor he couldn't come. Cauchois took it well, and later, Roski and Munson would pay Cauchois a few thousand dollars for his interest in the champion wedge, and Cauchois would go off and create his own robot.

The next day, Roski, Thorpe, and the rest headed to Germany: Frankfurt, Berlin, Hamburg, Essen, Stuttgart. Seven cities in two weeks, with a four-day break in the middle, riding in a luxury bus across the German plain with the disc jockey DB, another Smile DJ named Steve Stoll, and a techno band from New York. And, in the bus's luggage compartment, boxes of robots, electronic toys—and heads of lettuce.

They started in Frankfurt. A sign outside the rave hall said "Marlboro Robot Attack" and was set on fire. Smile employees handed out little boxes of cigarettes. Large, SRL-like theatrical robots from the German robo-art group BBM milled around the dance hall at the start of the festivities. Later, during their 20-minute show, Munson and another Smile employee donned the orange reflective vests and feigned swings of sledgehammers at the scooting lettuce heads. Then large white balloons were thrown into the ring, and La Machine was sent out to hunt them down. For the finale, DooMore and La Machine circled each other, pretending to spar under the heavy spotlights. The German ravers milled around, watching. They seemed bemused at first, then impatient.

Since the shows went late and took hours to clean up, the performers got little sleep on the trip, most of it on the bus. By Stuttgart, LaValley was getting quite good at directing DooMore to knock over the toy robots with its pneumatic punch. In Berlin, the builders met with Thorpe and demanded to put an end to the lettuce hammering; it was just too ridiculous. By Hamburg, things were dissolving. They hadn't had a proper night's sleep in days, Thorpe's performance art ideas were growing repetitive, and they were all getting sick of each other. In each city, they could feel the reservoir of audience enthusiasm diminishing as the night wore on.

During the preparations that day in Hamburg, Roski and LaValley wandered outside behind the club to smoke. Munson closed the heavy steel door after them, unaware (he swore) that it locked automatically. Roski and LaValley were trapped in the cold air of the German winter. Inside, the pounding techno music drowned out their rapping at the door, and then everyone was too busy watching the TV cameras gathering around Marc Thorpe to notice the ruckus coming from outside.

While Thorpe was about to be interviewed by various German news stations about the sport of robotic combat, outside, Roski was angry and frustrated, and continued to pound on the door. He usually got what he wanted, and what he wanted now was to get back inside, where it was warm. Roski also had one heck of a temper. He pounded on the door and when no one answered, he became incensed. LaValley stepped back in alarm as Roski kept pounding and, finally, jumped up and punched a glass window above the door. It shattered, jagged pieces of glass crashing down into the club.

Someone ran over and opened the door.

In a fury, Roski charged over to his cousin and accused him of deliberately locking him out—just as German TV producers asked Marc Thorpe The Question, the one that all journalists asked: "Is the sport of robotic combat violent?"

"Well, there's a creative impulse tied to destruction that is fascinating," Thorpe answered this time. "When it gets into an area where it harms people, then it crosses the line. But Robot Wars is a very healthy sport—"

"I'M GOING TO KILL YOU, YOU SON OF A BITCH! LET'S GO RIGHT NOW, OUTSIDE!!"

"—It sets an excellent example for conflict resolution without bloodshed—"

"TREY, IT WAS AN ACCIDENT, COME ON!"

"—and without being mean. It's better to resolve conflict with robots. Why have people's lives at stake when you can send in robots? Why send an astronaut up into space when you can control a rocket?"

"FUCK YOU, GREG. BRING IT ON! LET'S GO!"

"Of course, there are circumstances . . . but, uh, the issue has always come up . . . there's no reason why national or, uh, international conflicts can't in the future be resolved with robots."

Finally the TV cameras were turned off and the cousins simmered down. Later, the group performed in front of a Hamburg crowd that seemed as if it had inherited the combined restlessness of all the other crowds that came before it. During the La Machine balloon hunt, the ravers stomped around impatiently. During DooMore's pursuit of the toy robots, the ravers booed and hissed. Fifteen minutes into the show, they were visibly unhappy.

Then it was time for DooMore and La Machine to perform their slow-motion waltz. The robots were brought out, and they proceeded to stage the fake fight. It was dreary and repetitive and no one's heart was in it anymore. Then, out of the darkness of the club, empty German beer bottles started arcing into the spotlight and crashing onto the floor. Glass scattered amid the robots.

This was how the German cigarette tour ended: with the theoretical question of the violence of Robot Wars, the threatened violence between the two cousins, and the real prospect of violence from an angry mob of German ravers.

R/C controllers in hand, the performers backed out of sight, and waited for the exhausting journey to end.

Marc Thorpe and Gary Pini quarreled about the little things. For Thorpe, planning the third annual Robot Wars wasn't just a job, and it wasn't really just about his family's financial security anymore either. Robot Wars was his progeny—a second child. This would be his legacy long after the Parkinson's took over, freezing his faculties and leaving him unable to function. He completely threw himself into the planning, right down to the last detail. Like any true, dyed-in-the-wool gearhead, he refused to compromise on his ideas, his conception of how the event should be staged.

Gary Pini wasn't a gearhead at all. He was a music lover and a businessman who, when a light bulb burned out in his New York City apartment, happily trotted over to the phone to call his superintendent. He didn't understand Thorpe's obsession with manifesting a mental image of the show right down to the last detail.

Thorpe, for instance, wanted to darken the long skylight windows of Herbst Pavilion at Fort Mason. Robot Wars needed a certain ambience, an attitude. It should be dark, gothic, and cool.

Pini thought this was too much of a hassle. Over the past two years, darkening the skylights was one of the biggest aggravations of the event. It involved tinfoil, duct tape, hydraulic lifts, and renting the arena out for a few extra days. But Thorpe insisted on it.

Thorpe also wanted yellow-and-black warning tape running along all the pneumatic flippers and around the inside of the arena. It just looked better. Pini thought to himself, Why bother? The robots were going to fight regardless, and no one would notice the damn tape, but Thorpe insisted. The arena was his canvas.

Gary Pini said, "His vision was one of total theatricality that required a dark room with spotlights and blasting music. The methodology to achieve this became better every year, and it did look good. But I wouldn't have done it."

Profile's Pini, DB, and Tracey Miller all worked hard, but Thorpe worked harder than anyone. "We would work for 12 or 14 hours and call it a night," Pini said. "Then Marc would stay up all night and adhere the reflective warning tape in some particular pattern that only he envisioned."

Pini preferred to sacrifice a little of that artistic vision. Thorpe would rather die than have his idea co-opted and diluted.

"I would watch as people would bring in these amazing creations, some so beautiful it would nearly bring me to tears," Thorpe said of the early Robot Wars events. "So much love, talent, sweat, money, brains, and sheer determination that people poured into these robots."

The weekend finally rolled around, August 16–18, 1996. There was as much press attention as ever, and the number of entries kept growing. From 49 robots in 1995, there were now 83. The idea seemed predestined for the commercial mainstream, the first new sport of the new century.

In the wake of the Blendo fiasco, Pini and Thorpe bumped the six feet of plastic shielding around the arena up to eight feet. Pini prayed that another destructive robot wouldn't come along to endanger the audience. Fortunately, the Colossal Pictures guys were busy with a movie production, and Blendo didn't compete that year.

But there were other new ideas, and upgraded old ones that enthralled the competitors and another packed audience. Robot Wars was becoming a model of the engineering ecosystem at work: One builder introduced a concept, and someone else picked it up and pushed it further. For instance, Gage Cauchois, exiled from Team La Machine, returned to his old *Next Step* videotape from 1994 and saw the ineffectual forklift Tiny Tim from the first event. He liked that design, two metal prongs that could get under a robot and pop it into the air. He built his own forklift, with the same powerful customized R/C speedboat starter motors he had used in La Machine. He called his new robot "Vlad the Impaler."

Donald Hutson started building his robot after seeing a news story on Robot Wars over the summer of 1996. Hutson was a robot engineer at the Machine Psychology Lab for the Neuroscience Institute in San Diego, a nonprofit research center that tries to understand higher brain functions and replicate them in computers. In his free time, he pursued all manner of gearhead endeavors, including building R/C gliders and racing motocross. Robotic combat "was another way to exploit that mentality," he said. "I was set back by how utterly entertaining it was to watch robots get destroyed and how each robot had its own strength and weaknesses." He built his new bot around the idea that its weapon should swivel, providing 360 degrees of offense and defense. The resulting bot, Taz, made from refurbished parts, had two wheelchair motors, aluminum armor, cambered wheels, and a turret-mounted head that could swing a 15-pound stainless-steel arm, smashing opponents or lifting them in the air.

Forty-nine-year-old Marin-based computer programmer Gary Cline entered a robot in the featherweight category, named Spunkey Monkey. On its top, the wedge-shaped bot carried a toy monkey wearing a red-and-white-striped shirt, a gold headband, and black boxing gloves. When the robot moved, the toy monkey boxed, drawing peals of laughter from the audience. Cline—who wore a red-and-white-striped shirt of his own—won the award for the Strangest Robot. He, too, was now hooked on the sport.

Will Wright, the SimCity guru and the maker of JulieBot, was also at it again. His new entry, My Little Pony, dropped anchors on the floor and then drove away, unspooling a line of double-sided sticky tape, meant to ensnare rivals. Wright also set up his daughter Cassidy and two of her friends with a featherweight clusterbot called Triple Redundancy. The robot was really three R/C cars from Radio Shack, connected to each other by spools of tape. As the three cars went their separate directions in the ring, the tape unspooled into a sticky net.

Dan Danknick, a thin, bald software engineer in his early thirties from Orange, California, was a spectator at Marc Thorpe's first two Robot Wars in San Francisco. He read about the inaugural competition in *Wired* and actually paid friends to drive down with him to watch the event. ("Don't blame me if it sucks," he told them.) By '96, he knew he wanted to compete. He spent eight months and $2,500 building a middleweight robot, Agamemnon. It used a two-horsepower gas engine to run twin carbide-tipped cutting blades, and a bottle of compressed nitrogen to power a titanium-tipped spike. The bot also carried a video camera behind a plexiglass shield, which transmitted a robot's-eye view of the match to Danknick's goggles. Agamemnon was named not after the Greek king, but for the Omega-class destroyer on the TV show *Babylon Five*. Yes, Danknick was a nerd.

But, proudly, so were all the other competitors. By the third year of Robot Wars, a larger community of like-minded individuals was beginning to form. They saw each other in 1994 and again in 1995, and by '96 there were familiar faces, old friends and collaborators. Most of them got along famously, because they shared the same general outlook on life and a love for mechanics.

Rivalries, of course, sprang up as well. Danknick anticipated and prepared for a confrontation with the infamous middleweight La Machine. He even kept a photograph of the wedge on the wall in his shop, and called the robot "geometrically boring" in an interview about Robot Wars in a local newspaper. The cousins from San Francisco, still enchanted with their modicum of media celebrity, frequently searched the Internet for articles that mentioned them, and found Danknick's article online. They e-mailed Danknick about the comment and got into a heated e-mail exchange, which Danknick hoped could be amicably settled in the ring.

Sadly, while Agamemnon dominated the middleweight category, it didn't fight the famous wedge. Trey Roski and Greg Munson had rebuilt and rearmored La Machine with money from the NewTek sponsorship. The extra weight bumped them into the heavyweight category, where it had a new and formidable nemesis: Biohazard.

Like Dan Danknick, Carlo Bertocchini had attended the first two competitions and was ready to test his mettle. Back in 1994, at a meeting of the San Francisco chapter of the Robot Society of America, where Marc Thorpe presented his idea for a robotic combat event, Bertocchini responded enthusiastically. He felt that he had been preparing for this his whole life. Even as a kid, he was obsessed with speed and driving. "I was doing four-wheel drifts in my Radio Flyer wagon while most kids my age were watching *Sesame Street,*" he said.

As a teenager, Bertocchini had a Dodge Charger and a souped-up Dodge Challenger. He rebuilt the engine on the Challenger, adding a high-flow carburetor, a high-performance intake manifold, ported heads, and gas shocks. Between the ages of 16 and 20, he spent most of his spare time and money on those two cars. Sleepy Belmont, California, didn't have many good streets for racing, so it was off to the hair-raising curves of the Woodside hills or, even better, Highway One, on the precarious rim of the West Coast, a long stretch of asphalt that can send even a cautious driver to an unscheduled meeting with his or her maker.

As a grown mechanical engineer and robot hobbyist, Bertocchini was already a champion in the biannual robotic sumo events held at the San Francisco Exploratorium, and had captained the team that won the annual FIRST robotics competition for high-schoolers and their mentors in 1995. Now he wanted to capture the triple crown of robotic events.

The first incarnation of Biohazard looked like a pizza box. It stood four and a half inches off the ground and, without the skirts that would come later, was perfectly rectangular. On its flat top, Bertocchini painted the medical symbol for biohazardous waste, in bright red. For a weapon, he conceived something new and formidable. He fashioned a magnesium flipping arm based on the mechanical engineering concept known as the four-bar linkage. At rest, the arm fit snugly into the robot's base—you couldn't even tell it was there. When the arm was deployed, it pushed forward and upward, flipping opponents over. Unlike Blendo, Biohazard wasn't going to destroy rivals, and unlike La Machine, it wasn't going to push them into the walls. Biohazard was conceived around the simple idea of incapacitating other robots by putting them onto their backs. Moreover, it was beautifully machined, a marvel of titanium, aluminum, and magnesium.

The pits at the third annual Robot Wars were a louder, larger echo of past years. Builders admired one another's work and traded parts and advice, while Steve Carsey, the young executive from the British-based Mentorn, roamed the pavilion with a video camera, doing further research for the planned Robot Wars TV show in the United Kingdom.

Attention in the pits quickly centered on the new heavyweight entry, Biohazard. Something about Carlo Bertocchini and his robot felt unusual to the other builders. Finally, Dan Danknick put his finger on it. "Carlo introduced himself to me on Friday and disappeared," he wrote in his website diary. "It wasn't until Sunday that I figured out why his pit area looked so weird: no tools. The guy was really ready."

In its first match, Biohazard fought Patrick Campbell's new robot, Terminal Frenzy, a barrel-shaped bot with wheelchair motors and a frenetic four-foot-long hammer on top that pounded back and forth. Terminal Frenzy hit Biohazard once and put a small dent in its pristine armor. Bertocchini responded by flipping Terminal Frenzy on its side, so it couldn't right itself. He graciously tried to flip it back over, but by then Campbell's weapon motor was fried.

Bertocchini next faced Gage Cauchois's new Vlad the Impaler. The two powerful bots pushed each other around the arena, until Biohazard nudged Vlad to the side of the ring, where it got entangled in a mousetrap. Vlad shook itself free but one of its wheels got stuck on top of Biohazard, and Vlad got taken for a ride around the ring. At the end of the match, Cauchois finally positioned Vlad's two forklift prongs under Biohazard and, activating the weapon, jolted it up into the air. Biohazard went up and over, and the crowd thought the match was over. But Bertocchini showed them the second use of his clever lifting arm: Like a trapped insect, Biohazard pushed against the ground, propping itself up and tipping over. The crowd had never seen a robot self-right before. It cheered vigorously and awarded Bertocchini the victory.

Setrakian's The Master, upgraded again for its third competition, went down fighting against the new robot Taz, and the upgraded La Machine routed the Tilfords' remodeled South Bay Mauler, sending it careening four feet off the ground. In their semifinal match, La Machine pounded Taz into a wall of the arena, ramming it so hard that the armor on its swivel head broke off, which left Bertocchini facing Trey Roski and Greg Munson for the heavyweight championship: La Machine versus Biohazard.

The cousins assumed they were on track to win another title. They had won the previous year, they had won in England, and now they were strutting about the pits like conquering heroes, doing interviews for overseas television news programs and answering the questions of every local print and TV outlet. "We were caught up in the frenzy," Munson recalled. But Bertocchini was gunning for them.

"Did you check the screws on the battery box?" Roski asked his cousin before the final match. Roski was worried that the damage from the performances in Germany might catch up with the wedge.

"Mmm, yes, I checked it," Munson lied.

Biohazard and La Machine met halfway across the arena. Roski tried to scoop Biohazard with its new curved plow, but Biohazard sat so low to the ground that the wedge rolled right over it. Then Bertocchini pivoted his robot and activated the sneaky lifting arm, which grabbed La Machine and hoisted it into the air. Roski's bot was now at the total mercy of its enemy. Underneath it, the battery box split open.

Biohazard dragged La Machine around the arena like a toy. Then it backed off and La Machine banged down onto the cement floor. Roski looked desperately down at his radio controller. It wasn't responding.

Biohazard grabbed La Machine again and hoisted it into the air, and the wedge's batteries plopped out of its bottom like animal droppings. Roski just stared in disbelief. The match was over. Bertocchini leaned over to Roski and told him, "I never lose."

After the knockout victory, the heavyweight title was awarded to Biohazard, and reporters clustered around the La Machine pit table for comment. Roski remained sulkily silent; Munson was more talkative. "I kind of feel like Miss America," he said. "You wear the crown for one year, then you have to give it up."

La Machine was in poor shape for the melee later that day. But in a story the cousins still enjoy telling, other participants migrated over to their pit table to help rebuild it. Peter Abrahamson replaced a pin in the motor. Tim Jenison of NewTek, La Machine's sponsor, produced a soldering torch and began to reconnect wires. Munson and his girlfriend repaired the battery box. It was a team effort.

And Roski got his revenge in the heavyweight melee: Working in tandem with their new friend, Scott LaValley and his robot DooMore, La Machine scooped up Biohazard and slammed it into the corner of the arena, bending its lifting arm. Unable to defend itself, Biohazard was at the mercy of the other robots, which pushed it into walls and ultimately flipped Biohazard on its back, where it couldn't revive itself with its crooked weapon. None of the other robots could put up a fight against La Machine. They each stopped moving, one by one, including The Master, whose saw popped off and, for a surreal twenty seconds, propelled itself around the arena on its own. La Machine was declared the winner. "We all lose sometime," Roski told Bertocchini after the melee.

After the event, Roski returned to Impact Media, and to thinking up ways to deepen his involvement in this sport. He worked on the Robot Wars website and edited a video of Robot Wars '96 for Marc Thorpe. He even edited various versions that focused solely on La Machine's triumphant victory in the melee, including one set to the tune of the Bee Gees' "Staying Alive." He took those tapes to The Forum in Los Angeles and played them on the jumbotron during L.A. Kings hockey games. The crowd appeared interested and amused.

Slaving away on customer accounts and trying to drum up new business, Roski's cousins, Greg Munson and Gar Moss, were beginning to resent Roski's obsession with robotic combat. "He was getting one-third of our revenue and not doing anything but thinking about Robot Wars," Munson said. "We were pissed."

Roski was also spending lots of time on the phone—talking to his new friend, Marc Thorpe. Thorpe was having problems with his business partners in New York. With his money, connections, and love for the sport, Roski thought he could help.

But to understand what happened next, and why Marc Thorpe's dream turned into a nightmare that would nearly destroy him, you need to know more about Steven Joel Plotnicki.

Chapter 4

Courtrooms and Other Arenas of Violent Combat

The joint venture agreement that cemented Thorpe and Plotnicki together was simple but fundamentally flawed. The agreement gave each partner exactly 50 percent ownership of the venture and also gave each partner the absolute right to block the other. Absent complete agreement, nothing could be lawfully accomplished. This was a recipe for disaster that was cooking from the moment the parties signed the agreement.

—Attorney William Weintraub, fee application before Santa Rosa bankruptcy court, May 2001

The rappers signed to Profile Records always figured Steve Plotnicki for just another wealthy white guy pulling the strings in the music business. So Plotnicki would take them out to the housing projects in Williamsburg, Brooklyn. That's where his mother grew up, where he spent the first year of his life. "They used to freak out," Plotnicki recalled. "They always imagined me as a posh guy with a silver spoon in my mouth. But I was just a another guy born in the housing projects."

Plotnicki, the music executive who bankrolled the first three Robot Wars events, was raised in a fairly Orthodox Jewish home. His family eventually moved to the Bayside section of Queens, a short bus ride from Shea Stadium and his beloved New York Mets. His father was a Polish Holocaust survivor who emigrated after the war; his mother dropped out of high school when she was young to help support her family after her father died of a heart attack. Plotnicki was their only child. In what he now considers an ironic foreshadowing of his future life as a frequent litigator, he attended Benjamin N. Cardozo public high school—named after the influential legal scholar and Hoover-appointed Supreme Court justice.

After graduation, Plotnicki attended Bronx Community College for all of six weeks before he dropped out to play rhythm guitar in a band called The Dice. As a full-time rocker, he toured colleges in the northeastern corridor, but after a few years he began to hate the fast and loose musicians' lifestyle and tried his hand as a songwriter. This was the late seventies, and even though Plotnicki was primarily interested in rock-and-roll and reggae, the recording industry at the time craved one thing in particular: disco tunes. Plotnicki and a friend, Elihu Rubin, composed a dance song called "Love Insurance," which they sold to MCA's Panorama label, where it was produced by another young music maverick named Cory Robbins. The song's treacly lyrics were finessed by the late disco singer Sharon Redd.

> Sometimes when you fall,
> you tumble so fast,
> there's no way to stop her from breaking your heart.
> You pick yourself up,
> but stumble again,
> and think it's the end,
> what you need is some love . . .
> insurance. . . .
> you're not alone,
> pick the phone I'll give you some love . . .
> insurance . . .
> I'll help you through,
> I'm waiting to insure love with you.

"Love Insurance" was a hit in the clubs and made it to the top five on the Billboard disco charts. Plotnicki got a check for a measly $3,000.

"It made me realize that you needed to be prolific in the way a writer like Marvin Hamlisch was to make any money, and I didn't see myself becoming that accomplished," he said. In 1981, 21-year-old Plotnicki and his producer, 23-year-old Cory Robbins, decided to set out in the music business on their own. They each borrowed $17,000 from their parents, moved into a cramped, one-room office in Midtown Manhattan, and called their new business Profile Records.

Profile made a name for itself by selling the new music form called rap. In the early eighties, rap artists and producers were selling their wares off the backs of trucks. Robbins and Plotnicki signed some of the first hip-hop artists in the business, and produced 12-inch vinyl records of such trail-breaking rap acts as Jeckyll and Hyde. The crown jewel of their label came along two years later, in 1983: Joseph "Run" Simmons, Darryl "D.M.C." McDaniels, and Jason "Jam Master Jay" Mizell, a trio from Hollis, Queens, made up the pioneering group known as Run-D.M.C. In their first two years, Run-D.M.C. sold three million albums and inspired a whole new wave of hip-hop artists. Suddenly, Robbins and Plotnicki were raking in profits and sitting on the cutting edge of an innovative genre capturing the attention of mainstream America. Even better, they were upsetting the apple cart of the major record labels, which had dismissed rap as a passing fad and were now missing out on the next big thing. For Profile Records in the early eighties, life was sweet.

But for the suddenly popular Run-D.M.C., it was increasingly bitter. The band faced the common paradox of success in the music business. When they signed with Profile for the bare-bones music industry minimum—including a paltry signing advance of $2,500—they had few alternatives. Nobody else quite understood rap music. But then their success helped establish the genre as a commercially viable art form. They were the hottest act in popular music, and they could now get a much better contract elsewhere.

Russell Simmons, Run's older brother, Run-D.M.C.'s manager, and the future founder of Def Jam Records, started looking for a better deal. He began a dialogue with Sony, which tried to negotiate with Profile to share the revenue on a new Run-D.M.C. contract. With the marketing might of Sony's CBS Records behind Run-D.M.C., they argued, everyone would benefit from a new deal. But

Plotnicki didn't like it; Profile had gambled on Run-D.M.C. and won big. Why should they share the bounty?

Frustrated with what he saw as Plotnicki's intransigence, Russell Simmons sued Profile in New York county court. Auditors and experts poured into Profile's offices to check the financial books and ensure that the band was receiving its due royalty.

Profile fought the lawsuit successfully, and Run-D.M.C. was ordered to pay all the legal fees in the dispute and lengthen its commitment to make records for Profile. After a year out of the limelight, the band released its third record, but musical tastes had changed. The trio was already fading into pop music history. Russell Simmons never forgave Plotnicki, whom he described to friends as the villain in Run-D.M.C.'s downfall. "Plotnicki alienated the guy who is one of the biggest entrepreneurs in pop culture over the past twenty years," says Nelson George, a preeminent hip-hop writer and the coauthor of Simmons's autobiography.

For his part, Plotnicki felt sorry for the members of the band. Here were three guys who should be happy with their success and the money they were making. But, he concluded, their egotistical manager, eager to start his own record label, manipulated them for his own ends and led them into disaster.

There were other lawsuits—this was the music business, after all. Like other music entrepreneurs, Plotnicki learned to be tough, aware of his rights, even paranoid about his business prospects amid the music giants that were jealous of what Profile owned. Big companies could behave like bullies. The little guy needed to look out for his own interests, and Steve Plotnicki learned how to do that.

One day in 1993, he walked into Robbins's office and told him he wanted to split up the business. He was no longer the musician who wrote "Love Insurance." He had successfully guided Profile's business matters for the last decade and a half, while Robbins ran the A&R side of the label. Plotnicki and Robbins had grown apart, and Plotnicki no longer trusted Robbins's creative judgment.

Robbins was hurt. "All we did was make money with Profile, millions of dollars," he said. "I still really don't know why he wanted to split up. I didn't want to end the partnership. It wasn't unenjoyable to me."

The two began to try to disentangle their intricate business relationship. Both scoured the music business, trying to raise enough cash to buy the other out. Things eventually grew cold, then altogether chilly. Plotnicki sat on one side of the office, and Robbins sat on the other. They hardly spoke.

The race to buy each other's share of the business produced such tension that in late 1993, Plotnicki moved out of Profile to the offices of Landmark Distributors. Landmark was a company Plotnicki had created to circumvent the monopolistic distribution network of the big labels. Instead of going through main channels and having profits squeezed, Profile shipped records to sister company Landmark and made money from that part of the music supply chain too.

From Landmark's offices in New Jersey, Plotnicki waged a long-distance battle for control of Profile Records. Employees at the time were subjected to *War of the Roses*–style warfare between the partners; Plotnicki cut off the flow of revenue from Landmark, so Profile couldn't pay its bills. No one got paid, and employees started hitting the job market. Robbins found a Landmark account with $250,000 in cash and moved it over to Profile, without Plotnicki's permission.

"The relationship between Mr. Robbins and Mr. Plotnicki is so fraught with tension over their inability to come to an agreement on selling their business," said a judge at the time, "that they mutually started putting economic pressure on one another."

Meanwhile, Profile's remaining employees holed up in their offices and whispered to each other like frightened children whose parents were fighting downstairs. They tacked random newspaper headlines to the wall, including one from the *New York Daily News* that read "Make it Stop!" and they glued little pictures of Robbins and Plotnicki onto the page.

One day, Robbins learned that Plotnicki was remotely accessing the Profile computer system. He picked up a pair of office scissors, marched down the hallway to the Profile computer room, and started cutting cables.

This was the music business, after all. It's a cutthroat world.

After a few weeks of battling, Robbins thought he finally had suitors in line to buy Profile. Tommy Boy Records, a division of megalith Warner Music, wanted to purchase the company for more than $15 million, and Robbins would stay on as president. But Robbins and Plotnicki couldn't agree on how to split the proceeds from the sale. Robbins was going to have a job, Plotnicki argued, while he wasn't, so he should get a bigger cut of the pie. In negotiations, Plotnicki was relentless. He never stopped angling for an advantage or resisting what he thought was an attempt by the giant record company to exploit Profile's wounded assets. In the end, the partners couldn't agree on how to divvy up the money, and the deal fell through. Plotnicki ended up buying Robbins out for $4 million.

Only a day after Plotnicki closed the deal with Robbins, the company's spurned suitor—Warner's Tommy Boy label—plus several other customers of Landmark filed an involuntary bankruptcy petition against the company in federal bankruptcy court in New Jersey. This was an all-out act of war—the business equivalent of an assassination attempt. The music industry publication *Variety* ran a major story that day, reporting that Landmark was insolvent and that it owed Tommy Boy and other customers more than $400,000. This was exactly the plan of Plotnicki's enemies: to scare Landmark's customers by drumming up bad publicity, strangling Profile's primary revenue source and making it vulnerable to a hostile takeover.

It was a realization of Plotnicki's worst fears: The big music giants were going to rip him off. He now knew who his true enemy was: the president of Tommy Boy, Profile's spurned suitor, a man named Tom Silverman.

Plotnicki countersued, charging that the whole thing was a grand conspiracy against him. He claimed that Cory Robbins, like the band members of Run-D.M.C., hadn't possessed the wherewithal or savvy to wage the battle for control. Tom Silverman, like Russell Simmons, was the more powerful force in the background, pulling the strings.

In dismissing the original bankruptcy petition and handing Plotnicki the first victory in the long case, Judge William F. Tuohey wrote that Tommy Boy and Tom Silverman "were frustrated that their quest to acquire Profile . . . ended in failure." The judge further found that the filing was not for "legitimate business reasons but was filed for vindictive motives to punish Plotnicki and Landmark for the breakdown in acquisition talks."

After several years of further fighting on several fronts in several different courts, Plotnicki won the war. In the grand settlement of the dispute in 1997, Profile Records was awarded more than $10 million in damages. Fighting in court had paid off: David, in the form of a small label known as Profile Records, had slain the mighty music industry goliath of Time Warner.

In the two major business disputes of the first half of his career, with Run-D.M.C. and Warner Records, Steve Plotnicki was stubborn, combative, and practically married to his lawyers. (In fact, his wife, Linda, was a lawyer, and a partner at one of the firms Plotnicki often employed.) The formula made him a rich man. Perhaps it was inevitable that these characteristics seeped into his other business dealings. In 1997, he ended up fighting in court with his business partner at Landmark, Burt Goldstein, when they couldn't agree how to divvy up the proceeds from the Time Warner settlement.

Plotnicki also fought with one of Landmark's customers, Ruff n Tumble Records. And right around the same time, Joseph "Run" Simmons of Run-D.M.C. declared bankruptcy, again trying to escape his seemingly endless contractual obligations to Profile. Plotnicki fought him, again, in Long Island bankruptcy court and in New York county court. As part of that case, Plotnicki tried to subpoena Simmons's phone records and credit card bills, to try to find out more about the private business relationship between Run and his brother/manager Russell Simmons. Joseph Simmons's bankruptcy case was eventually dropped—another victory for Plotnicki.

"I've been in the entertainment business a long time," Plotnicki explained. "What I do is finance people's ideas who have no money and can't find anyone else to finance them. When the business idea becomes lucrative, they don't need the money anymore. They are happy to take a better deal for themselves. Well, they don't like someone saying 'no' to them."

"It's how it comes out in the end that really counts," Plotnicki continued. "I've been the villain with recording groups who wanted to go to Columbia and Warner. It's part and parcel being an independent businessman in the entertainment industry who finances other people's ideas, where people need money because they can't finance their ideas in other ways. The consequence is that people don't like you when you hold your ground."

It's true: Many people in the music business don't like him, though they often express a grudging admiration for his tactics. They say he has a unique willingness to spend the time and the money to defeat his enemies in court; that he has a complete familiarity with the tolerance points of the legal system, and an inexhaustible patience for outlasting his enemies in protracted litigation.

"He should have been an attorney," said Doug Bail, the former chief financial officer of Profile Records, who also skirmished with Plotnicki in court. "He would have been one of the richest attorneys in the world. Every lawyer I've talked to is amazed at the strategy that he puts together and how he uses the legal system to his advantage."

In discussing his ongoing litigation, Burt Goldstein called his dealings with Plotnicki "the greatest swindle of my life." Goldstein and Plotnicki were pals before they were partners in Landmark and then enemies in court. He recalled going to the movies with Plotnicki and their wives and walking down the theater aisle. Plotnicki would ask, "Where do you want to sit?" Goldstein would say, "Well, let's sit there." Plotnicki would say, "No, I don't want to sit there. I want to sit here."

"It was always going to be his decision," Goldstein said. "Even after he tells you that it's yours."

"I can't argue with you, I never win," Goldstein said he once told Plotnicki.

"My wife says the same thing," Plotnicki responded.

"Steve is different," said Gary Pini, who worked for Profile for more than 15 years. "He won't let people do whatever they want and sometimes that involves getting lawyers and suing people."

During more than two decades in the music business, Steve Plotnicki never lost a major legal battle. He undoubtedly put many of the children of his attorneys through good colleges. A review of the court documents stemming from his many fights suggests that huge swaths of rainforest have been cleared to produce the paper for the voluminous briefs in his cases.

In 1994, Marc Thorpe knew nothing about his real partner. He thought he would be working with the affable Gary Pini. He was desperate for money to realize his idea and completely naïve when it came to business. He didn't check the history of Profile Records at all.

For the first few years, Plotnicki thought of Robot Wars like a new band. He referred to it as a property, not a competition, or a sport. He thought there was little economic value in the idea itself. What was more valuable was what Profile would provide—the marketing muscle to turn it into a television show, and then a line of toys. "You create a concept, you create a name. They have some value. But either you know how to make it commercially successful, or you don't," Plotnicki said. To the music exec, Profile itself contributed the value to the Robot Wars concept. He allotted little to Thorpe's side of the equation—the creative decisions, and Thorpe's tireless, around-the-clock work on the annual events.

Coming into their partnership, Thorpe and Plotnicki were similar only in one way: They both craved control. Thorpe, who possessed the gearhead sensibility, desired authority over the sport he had created and nurtured, right down to the reflecting tape on the arena hazards; Plotnicki, who possessed a super-charged, music industry ruthlessness, never ceded control of his business matters to anyone. The joint venture agreement, crafted that first year, delineated the partnership at 50–50. Neither could make a decision without the other. By signing it, Marc Thorpe locked himself into a room with Steve Plotnicki and threw away the key.

The first three years of the Robot Wars partnership were comparatively uneventful. Disagreements over the darkening of the skylights at Fort Mason would later be reflected upon fondly.

Steve Plotnicki attended the early Robot Wars events and brought his twin boys. They opened his eyes to the potential of the property. "It was their general reaction that convinced me," he said. "Children between the ages of eight and 12 seemed to be enamored with the robots. They wanted to go backstage to the pits and see them. That's what made me think it had a commercial future."

Like his employee Gary Pini, Plotnicki was not a gearhead. As a child, Plotnicki watched his peers spend their afternoons putting together crystal radios from kits. He could never understand why they seemed so entranced; he was more interested in playing baseball or taking the bus to Shea Stadium to watch the Mets.

In the first few years of Robot Wars, Plotnicki didn't even pay much attention to the business. He was sidetracked with the battles with Warner Records and Run-D.M.C., and it was Gary Pini's project anyway.

Still, he never intended to let that joint venture agreement govern the business over the long term. He wanted to institute a more comprehensive limited liability corporation, which would enjoy tax advantages in the state of New York.

But by the time the second Robot Wars rolled around, Profile's loans to the Robot Wars partnership had surpassed $200,000. Its first $50,000 had bought a 50 percent stake in the company and now Plotnicki wanted a bigger percentage of this new LLC. More money, according to the formula for financing start-up businesses, buys a greater interest in the company. He felt Marc Thorpe surely knew about this business practice and had agreed to it.

But Marc Thorpe hadn't agreed to it and wouldn't accept it. Thorpe felt he had signed up to remain an equal shareholder. For Thorpe, taking a minority share meant losing control. Robot Wars was his brain-child and if he was unwilling to compromise on the blackening of the windows at Fort Mason, he was surely not going to hand over his influence over the entire direction of the sport. If there was a meeting over a toy deal, he wanted to be there to explain to the licensee what the whole thing was about. Mentorn Films in the United Kingdom, for example, was planning its very own Robot Wars TV show; Thorpe wanted to consult on it, and help Mentorn execs Tom Gutteridge and Steve Carsey manifest this part of his vision.

When it came time to revalue the partnership, Thorpe argued, "I put in at least as much value in my time and labor as they did with their money." Negotiations over turning the partnership into an LLC went nowhere in 1994 and '95.

Robot Wars '95 came and went. Sometime that year, Steve Plotnicki began to pay closer attention. Why was his company throwing money into this entity called Robot Wars for nothing, he asked, not a bigger percentage of the company, and certainly not a profit, for Robot Wars was losing money each year? He started to look closely at the business arrangements, and noticed that Profile was paying Marc Thorpe a salary of $1,000 a week—something not required in the original joint venture agreement. Maybe, Plotnicki thought, confronting Thorpe with the possibility of losing his salary would force him to reconsider his stance.

Plotnicki started putting financial pressure on Thorpe, via Gary Pini, the go-between. "It can't go on like this anymore," Pini told Thorpe. "You have to get outside employment. We can't keep paying you if no money is coming in."

Thorpe couldn't just go out and get another job. He was putting all of his already-limited physical energies into Robot Wars. He didn't want another job, plus who would hire him? He would sit in another office working on something else and still be obsessed all day about the next competition.

The outline of a standoff emerged: Thorpe would not agree to any final form of the partnership that sacrificed his veto power over creative decisions, or that gave Profile a majority percentage of the company. Plotnicki would not keep pouring money into Robot Wars and paying Thorpe a salary without renegotiating their business relationship to give him definitive control.

In May 1996, after the Marlboro cigarette tour of Germany, Gary Pini told Thorpe his salary was being discontinued. "Go get a job, Marc," Gary Pini said.

Thorpe stopped getting paid but kept working on Robot Wars anyway. With his family, Dennie and his daughter Megan, he lived off credit cards and personal loans from friends. He was sure his vision would eventually pay off. There were whiffs of interest in licensing the Robot Wars concept from toy makers like Hasbro and Tyco, and video game makers like Virgin Interactive, which offered to pay the venture $50,000 for a three-year option to make a Robot Wars video game (the game company would later go under). Plus, after two years of planning their TV show, Mentorn was getting serious interest from the BBC in creating six pilot episodes in London. If that happened, royalties would start flowing in. He just needed to hold on, so he continued to run the Robot Wars business full-time from his home office, answering phones, faxes, and e-mails, paying bills and planning ahead.

Robot Wars '96 was a big hit among the builders and the press, but for Plotnicki, it was just more money down the drain. His loans to Robot Wars now totaled $354,000. He called Pini, DB, and Tracey Miller into his office and said, "Enough! Don't spend any more money on Robot Wars. I know you all love it. But it's my money. I won't do anything more until this is resolved."

The lawyers continued the exchange of increasingly heated letters that year. Profile's lawyer suggested they create the new company with Plotnicki holding 45 percent, Thorpe holding 44, and Pini and DB both at 5.5. They also insisted that Thorpe find other employment.

Thorpe's lawyer took a month to respond and, when he did, rejected the proposal. "It is necessary to find a method by which Marc can be relieved of the worries that he has in being a minority shareholder," he wrote. Instead, Thorpe's lawyer proposed that Thorpe sell a small percentage of his shares to Plotnicki for an additional $50,000. Plotnicki would then have the majority of shares, and Thorpe could alleviate his financial problems. But Thorpe insisted that Profile employees all provide written assurance that he would maintain creative control over all aspects of the event production. It was a minor compromise.

Plotnicki wouldn't agree to that. A licensee like Mentorn in the United Kingdom had to be free to take Robot Wars in its own direction, he said. They were the experts on television—not Marc Thorpe.

As the standoff simmered, Thorpe's new ideas for expanding the Robot Wars brand were rejected. He wanted, for example, to print and sell a line of robot trading cards. He brought the idea to Gary Pini, who brought it to Plotnicki, who said it was too expensive. Later, Thorpe would charge that Plotnicki was rejecting everything, including offers from TV and toy companies, because he didn't want money coming in before the partnership was renegotiated along more favorable lines.

Finally, Gary Pini got elbowed aside. Plotnicki decided to talk directly to Thorpe and get this over with. In court documents, Thorpe recalled the typical conversations this way:

"How much do you want, Marc? How much do you want for 30 percent? I want you to think about what you would accept for a certain percentage."

"I'm not going to give up my share of control, Steve."

"How much do you want? I want your veto! How much do you want? Tell me how much it's going to cost!" Conversation like this ended with Thorpe holding the phone about three inches away from his head, as Plotnicki vented his frustration.

Thorpe was stubborn; Plotnicki was belligerent. Positions hardened, lines were drawn in the proverbial sand, harsh things may or may not have been said in heated moments. In a conference call with attorneys from both sides in the spring of '97, Plotnicki allegedly said, "I will pursue him past bankruptcy and destroy him and his family." He later denied having said it and both he and his lawyer filed affidavits claiming those words were never spoken. But Thorpe, at least, believed it had been said, and it colored his attitude and all his future actions. He considered it a legitimate threat. He lost faith in ever being able to work out a solution with his partner at Profile Records.

As 1997 progressed, things looked bleak. Marc Thorpe educated himself on the finer points of contract negotiations and learned enough to know that he didn't know anything. Pini had once suggested Thorpe go looking for new Robot Wars investors, so he found a sympathetic ear in Bob Leppo, a San Francisco–based venture capitalist. Leppo, 58, was a self-described speculator and stock market poker player: He had taken risky bets in the eighties on rocket entrepreneurs like Gary Hudson, who wanted to send private citizens into space. Hudson's first rocket, the Percheron, exploded on the test stand. Leppo did better with software companies like Advent, and in the nineties, with a successful web hosting company, Best Internet.

Thorpe showed the avuncular Leppo a videotape from the early events. Even though he was "aggressively clueless" when it came to all matters mechanical, Leppo thought the events showed unmistakable promise. "The audience participation was electrifying," he said. Plus, he noted, "When something weird comes along, I'm interested."

Leppo agreed to invest. He held a conference call with Thorpe and Plotnicki to try to work out the details, and quickly found things weren't that easy. Plotnicki insisted that any percentage of the company that Leppo bought would come directly from Thorpe's share. Plotnicki didn't want a third partner, ostensibly an ally of Thorpe, who could act as a tie-breaker. "It was another way of saying the joint venture didn't want my money," Leppo said.

Instead of buying some of Thorpe's shares, Leppo did the next best thing: He ended up merely giving Thorpe a loan, which Thorpe used to hire a new attorney and to prepare for the next event, which Profile had said it wouldn't finance until the parties found a resolution to the impasse.

As the summer of 1997 approached, the partners moved closer to working out a deal. Profile resubmitted a plan that would break the company up among Plotnicki, Thorpe, Pini, and DB. They sat back and awaited comments from Thorpe. Great gusts of cold air were howling over the robotic combat landscape.

What did Profile know about Robot Wars 1997, and when did they know it? Plotnicki and Gary Pini said they were caught completely off guard. However, Tracey Miller, who had left Profile to start her own PR agency but kept in constant touch with Profile, said she knew about it all along and doubted her former colleagues could have truly been in the dark. Marc Thorpe insisted it could not have been a surprise.

Nevertheless, in July 1997, with the business dispute still unresolved, Gary Pini said he suspected something was amiss. Thorpe seemed in no hurry to talk about preparations for the fourth event. Pini picked up the phone and called the Bass ticket outlet in San Francisco, and asked if there were any tickets available for Robot Wars '97 over the weekend of August 19 and 20. There were. He did a little bit more digging and found that Thorpe had also bought insurance, rented the hall at Fort Mason, and solicited the services of a production company to provide lights and sound.

Pini took this new information to Plotnicki, who considered it a violation of their joint venture agreement, and an act of insurrection. The agreement stated all decisions in the Robot Wars venture were to be made jointly, by both parties.

Other men would have handled the situation differently. Some might have sent their minions to befoul the enemy's car, or to padlock the gate of his home. Others still might have been content to settle the matter in the arena with combat robots, a battle between mechanically inclined minds. Steve Plotnicki was not that kind of guy. No, he had his own approach: He slapped Marc Thorpe with a $5-million lawsuit in the Federal Court for the Southern District of New York.

Peter Abrahamson, builder of Gigan and Gigan II, wrote the judge who would consider the injunction. So did Robert Orr, creator of the lightweight Nezumi, and Tony Buchignani, builder of featherweight Wedge of Doom, who complained that he had spent over $2,000 for parts and untold hours working on a bot that now wasn't going to get to fight. Mike Winter, the Wisconsin designer of X2, complained, "This is the largest project my daughter and I have ever worked on together." Gary Cline, of Spunkey Monkey fame, pegged his investment in a letter to the judge at $1,500, plus 800 hours of labor, and added that his brother had already bought airline tickets from Orange County to fly in for the show.

Dan Danknick, builder of Agamemnon, wrote a letter that began, in a big font, with the number $31,247. "This is the amount of money I will stand to lose if you cancel or delay the Robot Wars 1997 competition in San Francisco. . . . The thought of having all this time and work resolve to zero would be truly horrible." He added, "As an incentive for you to yield, I've enclosed this new crisp one-dollar bill. You may think of it as my compensation to you for reading this letter. Spend it as you wish, perhaps brightening the day of a child."

The gearheads were angry. They were not yet a unified community—they only met once a year in San Francisco—but they were articulating a single goal: They wanted to compete. A Profile Records lawyer had posted on the new Robot Wars website, erected by Trey Roski, that the event might be canceled due to legal troubles. Thorpe responded with an e-mail to all participants: "For the record, I have done nothing wrong and believe that Robot Wars '97 has every right to go forward." Then he gave participants the address of the New York court and implored them to write.

It's unclear if, and how closely, federal Judge Deborah Batts reviewed the dozens of letters that poured in.

Batts called the parties into court in early August 1997—less than two weeks before the fourth Robot Wars event was supposed to be staged—and allowed the parties to vent.

Addressing the alleged "bankruptcy" comment by Plotnicki, Thorpe said, "Those intentions are so hostile and so upsetting and dangerous to me, that I've lost all trust and faith and confidence in my partner."

"If that's the case, both of you are in trouble," Batts said. "Because I don't see, based on the review I have made, either one of you walking out of here with the spoils. You have a valid joint venture agreement."

"If this year's event is a financial failure," Plotnicki said, "this really goes to the heart of the dispute and the heart of the economic problem, the parties' disagreement on how to hold this event and make it profitable. And Mr. Thorpe insisting his vision is the only way to do this, and I am required to pay the bill. I am giving him a blank check, you know, to really do what he wants. I am really trying to stop that part of the problem. That's why we're here."

Batts agreed, $368,000 in four years was a lot. Profile Records needed protection for its investment: If Plotnicki was going to put money in, he should get something in return.

Thorpe's new lawyer spoke up. "Yes, they are the wallet. Yes, they put in substantial money. Don't deny that for a moment. But they want to take away Mr. Thorpe's control, and Mr. Thorpe says, absolutely not. . . . He just wants the creative control. He will leave the financial [part] to them. Mr. Thorpe is not a businessman. Mr. Thorpe is a toy genius. He looks at the world different from the way you and I look at the world, which is always a bit unusual."

Thorpe's attorney gave the court an example. Marc Thorpe, he said, should be able to have creative input on any TV show based on a Robot Wars license.

"I can't control how much [a show's producers] feel they need him," Plotnicki interjected.

Batts switched over to Thorpe's side. "You can't give them carte blanche to take Mr. Thorpe's idea and turn it into something that is totally foreign, harmful, or denigrating to the concept that he has developed, for them to make money."

Thorpe's lawyer then drew attention to the noncompetition agreement in the most recent version of the LLC papers Profile was advocating. The clause said that Thorpe's right to stage live, local events in San Francisco should not conflict with any agreements that the company entered into with a licensee. This constituted one of Thorpe's greatest fears: Profile could license the idea to someone else, who would then decide to put on their own event in America, which would prevent Thorpe from putting on the competition himself. He could be totally pushed out of the sport he created!

Judge Batts had heard enough. "This is ridiculous. See, Mr. Thorpe came up with this idea. His idea has value. He has the right to be able to protect the integrity and also to protect those things that are important to him. You can't license away everything and shrug your shoulders and say, it's not my fault. . . . I also understand Mr. Plotnicki needs some protection. He's not a bottomless pit, he doesn't have a private road into the bank vault at Citibank or something else like that."

She commanded, "I want you to go to lunch, together. I want you to talk. You both have very valid propositions. But that and a subway token, without more, is going to get you out of this courtroom and up to Midtown together. You have to work with each other."

Thorpe looked at Plotnicki, and Plotnicki looked back at Thorpe. Already there was too much water under the proverbial bridge. Thorpe tried to escape and referred to Plotnicki's bankruptcy comment. He believed the comment was a warning that implied that as long as Thorpe tried to fight, Plotnicki would outmatch him in court with his superior resources. "What I am suggesting to you," the artist begged the judge, "if I can use a harsh metaphor, is that it's very difficult to conduct business with someone who holds a gun to your head."

Batts was no longer listening. She ordered the bailiff to bring in sandwiches, and Thorpe, Plotnicki, and his lawyers retreated to the jury room.

There's no record of the heated 120-minute discussion that took place that afternoon, but the parties described it as emotional and angry. Plotnicki and Thorpe yelled at each other, both lost their voices, and the sandwiches went uneaten. They came back into the judge's courtroom hours later, temporarily mollified.

They had worked out an interim agreement, just for that year's event. Profile sold Thorpe a license for one dollar to stage the fourth Robot Wars. There was also a blueprint for solidifying the business into an LLC, with each party holding a 50 percent share. But all that would have to be given final form in writing later.

Thorpe thought he had won. Plotnicki had brought them to the brink, then seemingly surrendered his demands to renegotiate ownership of the business. That's why Robot Wars '97 took place at all.

The robot builders returned to Herbst Pavilion and Fort Mason, on the edge of the San Francisco Bay, and they were jubilant. Based on the little that most of them knew about the legal dispute, it seemed that the whole thing was over and that Marc Thorpe, the good guy, had triumphed. Corroborating that impression was the fact that Gary Pini and DB weren't helping to run the tournament this year. Since Profile wasn't bankrolling the event, just licensing it for the year, Thorpe was now running the whole show, with backing from Bob Leppo.

To help with logistics at the fourth annual Robot Wars, Thorpe asked two friends, brothers Joel and Jim Hodgson, to be his production coordinators. Joel Hodgson was the creator of the cult cable hit *Mystery Science Theater 3000,* a sort of cultural metacommentary in which a man and two robots watch bad science-fiction movies and trade quips about the dreadful flicks. The Hodgsons' sophisticated, robot-inclusive humor dovetailed perfectly with the Robot Wars vibe.

With their assistance, the 1997 event was bigger and larger than ever. There were 91 robots, up from 83 the year before. This time, on top of the eight feet of quarter-inch polycarbonate shielding, two more feet of protective netting was added. The crocodiles would be hard-pressed to escape their pen this time. Thorpe also added a few hundred more seats—which meant he would be able to pay Bob Leppo back part of his loan.

The hazards in the arena were also different. The bowling ball and mousetraps were gone. Thorpe designed new dramatic stainless-steel pincers that pounded the cement on both sides at the center of the arena, and at each end, pneumatic bumpers that pushed back and forth. A robot that wandered over those spots could find itself in a heap of trouble. Blendo creator Jamie Hyneman and his special-effects colleagues fabricated and helped install the new arena features.

Trey Roski and Greg Munson had gotten ready for the event by transferring their robot operations to Scott LaValley's workshop at his parents' house in Novato, California. Applying his talent for bull-headed persuasion, Roski scored another $25,000 in corporate sponsorship from NewTek and other companies. With that money, he contributed to the purchase of a brand-new industrial-grade milling machine for the shop. LaValley, in addition to building his third robot, DooAll, helped Roski and Munson install a pneumatic battering ram in the face of La Machine's plow. This year, the team vowed, they were going to be prepared for heavyweight champion Biohazard.

Carlo Bertocchini had upgraded his robot as well. Though it had dominated in the face-off competition in 1996, La Machine had won the melee round. Bertocchini vowed that Biohazard wasn't going to lose to the same robot, in the same way, ever again. He added a "skirt" to his robot, hinged pieces of titanium angled out from the bot's rectangular frame on all sides, stretching right to the ground, so absolutely nothing could sneak underneath the robot's frame. He also upgraded parts of the lifting arm from magnesium to titanium and moved from four-wheel drive to six. "There's not much they can damage. Even the motors have hardened cases," he told *Wired,* which was working on a major Robot Wars story.

Other returning veterans and new competitors raised the bar in distinct ways. Second-year competitor Dan Danknick brought The Alexander, a 100-pound, six-wheel bot that had a milling cutter at the tip of an arm that rotated and flexed like the ladder on a fire truck, and could lower onto enemies from above. Childhood friends Alex Rose and Reason Bradley, two new competitors from Sausalito, brought a heavyweight robot called Rhino Halon. The bot didn't have a ramp, a pneumatic punch, or a spinning blade—it sprayed Halon, a gas used in fire extinguishers to snuff out gas fires and kill gas engines. And Gary Cline, who had cracked up the audience the previous year with the boxing Spunkey Monkey, this year brought Mad Monkey—another robot with a pugilist simian.

These were some of the interesting robots in 1997, but many others were boring wedges, inspired by the success of La Machine, and Thorpe was still worried about that trend. He wanted to nudge the builders as far away from conventional designs as possible. That would be the key to keeping audiences interested and expanding the Robot Wars concept. That year, he began to manipulate the rules, to encourage more variation in the kinds of robots that were being built. Instead of an audience vote, he instituted a panel of three judges that would determine the winner if both robots were still moving after five minutes. They ruled on damage, aggression, and control—with a tacit preference for more sophisticated robots. One of those judges, incidentally, was Jef Raskin, the *Wired* writer who had scripted the original article on Robot Wars.

A second change in the rules prohibited pinning opponents to the arena walls; many robots, particularly the wedges like La Machine, were using that as a technique to win fights. A third change added a weight advantage for walking robots. Hoping to inspire variety, he decreed that robots without wheels could enter the weight class below their official weight—giving them up to a 150-pound advantage.

Thorpe laughed at the memory of what happened next. "Then Mark Setrakian shows up with a snake."

The special-effects designer from L.A. had tired of The Master. Working on the set of the original *Men in Black,* Setrakian set out to take advantage of a long, snakelike handheld controller he was using to puppet one of the animatronic aliens in the film. He essentially built a larger version of the controller: This larger snake would exactly mirror the movements of the smaller controller. The Snake was composed of nine separate segments, each with its own radio receiver and power supply. It spanned 13 feet in length, and used 17 linear actuators to writhe, crash, and slither around the arena. Setrakian could make it move like a real snake, even curling its three-pronged claw mouth toward the sky. The Snake was breathtaking to behold, but it was not a true fighter. More elemental robots like the Scorpion, with a giant hydraulic spike, made a beeline for its defenseless middle, pounding it until the aluminum plating came off. Still, the crowd and competitors were overwhelmed by Setrakian's imagination and skill.

The crowd had something else to cheer for too—the return of Blendo. Absence supposedly makes the heart grow fonder, and with combat robots, it allows their reputation for carnage to grow to mythic proportions. By getting disqualified in '95 and then not showing up in '96, Jamie Hyneman had deprived his fellow competitors of seeing Blendo lose, and many of the robot builders concluded that Blendo had no weaknesses. "Most people felt that if Jamie lost radio control over Blendo, it was going to eat its way through the bleachers and go across Lombard Street," said Marc Thorpe.

The year before, Hyneman had left Colossal Pictures and founded M5, his own shop. Among other projects, he designed the animatronics puppet for the "'Lil Penny" Nike ads. When he brought Blendo into the pits that year, the entire pavilion got quiet. Once again, he wore fatigues and his ever-present black beret. ("I'm bald, and my head gets cold," he admitted.) Hyneman learned that for his first match, he was going up against Hercules, a double-sided wedge created by first-year participant Jim Smentkowski, an ILM employee from Marin County.

Smentkowski had been a spectator at the '96 Robot Wars and had immediately raced home to prepare his creation for the next event. He spent $2,500 and nine months' worth of weekends creating Hercules. Shock set in, he recalled, when he realized that for his first match, he faced the infamous spinner. He added a hammer attachment to his bot and rolled Hercules into the arena with a heavy heart, "Thinking that I may need to find a wheelbarrow to pick up the pieces afterward." After that weekend, he wrote about the match on his website:

> The whole thing lasted about 30 seconds. Blendo sat across the arena, spinning at what seemed like a million miles an hour, and I decided that my strategy of hitting him with the hammer just wasn't going to work. The only way to get him would be to take one hit from him, wait for him to slow down and try to flip him over. So, I went in for my first hit.
>
> The first impact sent one of my two-pound stainless-steel armor plates flying into the arena plexiglass wall and left a four-inch gash. . . . Had the wall been lower or less durable, someone surely would have been decapitated. Unfortunately, the impact did not slow Blendo down much, so I had to get hit again.
>
> I moved in and the second hit caught me on the right side corner and picked up my 169-pound robot and threw it into the arena wall. The impact warped the side of the case and pinched the motor wires on one side, which in turn shorted out my Vantec speed control. That was all she wrote, Hercules was dead. I had no control or movement. I rolled him away afterward thinking that perhaps this whole thing was a bad idea.

Smentkowski would have to install another $500 speed controller to get Hercules up and running again to fight its next match. Jamie Hyneman and his crew were thrilled.

Without Gary Pini present, it was Joel Hodgson's turn to start worrying. Could Blendo hurt a spectator? For Blendo's next match, Hodgson started looking for an opponent with smooth armor and none of the small external parts that might come loose and ricochet over the 10-foot barrier. He settled on Punjar, a wedge similar in design to La Machine, made of welded diamond-plate aluminum, built by San Francisco carpenter Ramiro Mallari.

Hyneman didn't like this at all. He felt Blendo's real power had yet to be tested and wanted a robot he could "spray" across the arena. Disappointed, he decided to embrace Blendo's image for destruction—to ham it up. His pit table was right across from Mallari's. "So I walked over there and asked him a few questions. Then I went back to my robot and started sharpening my blades, looking at him and grinning."

According to Hyneman, the tactic of intimidation worked. Psyched out, Mallari decided to shoot Punjar across the ring in a damn-the-torpedoes, all-out attack. The match lasted less than 20 seconds. Punjar raced across the arena at full speed and was launched about six feet backward, spinning like a Frisbee. The welds along its front cracked open and the electronics poured out. Punjar was finished. The crowd went wild for Blendo, the whirling-dervish metallic saucer.

Joel Hodgson had seen enough. He walked over to Jamie Hyneman's pit table. "Is Blendo capable of throwing another robot out of the arena?" he asked.

"Oh, it's totally possible," Hyneman bragged.

Hodgson conferred with his brother Jim and Marc Thorpe. "Once someone gets hurt in a performance, it changes everything," he said later, recalling the tough decision. "I just felt like there was a really big gap between the robot guys' love for their robot's abilities, and the danger of an innocent person getting hurt."

Joel Hodgson played the role formerly held by Gary Pini in the same production two years before. "You have a really great, impressive robot," he told Hyneman. "But due to safety concerns, we're going to have to ask you to leave the competition and accept the cochampionship of the heavyweight division."

The crowd booed when the decision was announced, and Jamie Hyneman raised his arms in a halfhearted celebration. The crowd wanted to see Blendo fight. Now, its vulnerabilities would have to wait to be exposed another day. That left the heavyweight class wide open.

After Blendo's disqualification, Biohazard faced Vlad the Impaler in a rematch of last year's epic fight. With its new titanium skirts, Biohazard's undercarriage was impenetrable to the two steel pipes of Vlad's forklift. Cauchois's robot kept driving right over the sneaky lifter and the two powerful robots jousted ineffectively for a few minutes, until Biohazard flipped Vlad the Impaler over. Then Cauchois triggered his robot's new self-righting mechanism, a single pneumatic cannon on the top of the robot that fired directly onto the floor. Vlad popped up and over onto its wheels, to a gasp of surprise from the audience. But Vlad couldn't do much else. At the end of five minutes, the judges awarded Biohazard a victory, 15 to 10.

Biohazard next faced Rhino Halon. Alex Rose's innovative gas weapon was working well and had immobilized Scorpion's engine in an earlier fight. Sadly for the engineers from Sausalito, however, Bertocchini used an electric engine in Biohazard. Rhino was quickly pushed against the wall and marooned on one of the hazards.

Biohazard than faced the refurbished South Bay Mauler. Supreme Commander Tilford had managed to augment the spinning Mauler's destructive power each year, but it never managed the deadly efficiency of Blendo. In this fight, Biohazard simply pushed Mauler into the wall, generating a tremendous crash. Directly behind the impact on the bleachers, a dozen spectators scampered away, fearing for their safety. Bertocchini had another victory, which left Biohazard, for the second year in a row, facing Trey Roski's La Machine in the finals.

Roski had been waiting and planning for this the whole year. He had spent long nights at the shop in Scott LaValley's home, honing his strategy for wreaking his vengeance on Bertocchini, who he felt was overly confident.

But in the ring, things didn't go as well as he planned. He sent La Machine into a charge, but midway across the arena, swerved to the left and got immediately hung up on the base of the arena hazard. Bertocchini swerved in right behind, deployed Biohazard's lifting arm and flipped La Machine onto its side. Game over. Time of knockout: 15 seconds.

Trey Roski just stared.

Afterward, reporters asked him how he felt. He didn't have a response. The heir to L.A.'s largest sports empire was numb. After a few hours, it began to sink in that he had spent so much time and energy and still lost to his nemesis in 15 humiliating seconds, and it was horrible—unthinkable. Finally, he appeased himself in the only way possible, with the justification of any athlete or fan who tastes the charred gristle of defeat: *There's always next year.* He would have his revenge next year, *they would get 'em next year!* And Trey Roski, who was learning how to accomplish the things he put his mind to, whatever the odds, resolved that it would be so. *Next year* he would climb back on top of the robotic combat world.

But next year never came. By then, the young world of robotic combat was totally frozen in a solid ice block of litigation, and it became Roski's goal not to win the game, but to save it.

Part II

Attack of the Human Money Bots

Chapter 5

The Ice Age

The main reason I build robots is that they're art. The fact that the robots can get destroyed brings even more power to the medium. I have a degree in art and a job in technology and have twice won at Robot Wars. I plan to continue building robots forever, and if Robot Wars isn't viable, I will put on my own competition. Most other combatants seem to have similar strong feelings.

—Robot builder Mike Winter, 1998

Stephen Felk was a gearhead; his living room gave it away. The 50-year-old cabinetmaker lived in a one-bedroom apartment in the Nob Hill neighborhood of San Francisco. The lathe, mill, drill press, band saw, table saw, and chop saw, the tools of the robot-fighting and carpentry trades, were wedged into his tiny, angular kitchen. There were a stove and a toaster somewhere in there too.

The living room was crammed with mechanical-themed books and magazines, as well as robot parts, screws, and every kind of metal and tool imaginable. They were piled high on the floor and on makeshift tables, an uproarious, unseemly mess. When guests stood in the room, they could hear the electronic warble of a fish tank, which was odd because there didn't seem to be—oh, there it was, underneath the pieces of aircraft aluminum.

Felk had thinning hair and ever-present bags under his eyes, evidence of long nights working on his combat robot, and he exuded intensity.

He grew up in Fair Oaks, Pennsylvania, a suburb of Pittsburgh.

His father was a Marine who landed on Guadalcanal during World War II, and later became a plumber. His dad died when he was eight, after a year-long battle with cancer that started in his right leg and went to his pancreas. That year was unspeakably awful, and young Felk, eager for a distraction, gave himself to mechanical absorptions at an early age. He visited a junkyard by a nearby creek and the gas station down the road, hauling back starter motors and other machine detritus, stockpiling it in the backyard, where it drove his mother and the neighbors nuts. There were also the treehouses: He built a new one every summer in the two apple trees behind his house.

Then Felk turned 11 and was old enough for the Soap Box Derby, also known as the "Gravity Grand Prix," the greatest amateur youth racing event in the world.

In the 1960s, Soap Box racing was wildly popular. Thousands of dads and their kids around North America built cigar-shaped, engineless cars. The kids cut the pockets off their jeans so they could squeeze inside the oval cockpits, and snug inside their creations, they rolled down gently sloping hills in one-on-one races. If a boy finished at the top of his local competition, he went to the Soap Box Derby national championship in Akron, Ohio, the tire capital of the world.

Felk stood out in the Soap Box circuit. He had no father standing behind him, designing the car or providing aid. Stephen built his cars alone.

Each year he started thinking about the race around Thanksgiving, when he'd begin to sketch a design. During the cold Pennsylvania winter and throughout the spring, he'd devote weekends and afternoons to the car: developing the chassis, testing the brake system, until the local race over the July 4 weekend. Felk was a formidable competitor and made it to the finals of his age class when he was 13 years old. He broke his arm when he was 14 and sat out the year, but returned at 15. That year the brakes failed on Felk's second run and the car smashed through the barrier at the bottom of the hill and wreaked havoc among the judges and spectators.

At 16, Stephen Felk was too old to compete anymore in the Soap Box Derby. He would spend the next 30 years trying to recapture that sense of community, creativity, and competition that loomed so large in his young imagination.

After high school, he hitchhiked to San Francisco and joined the burgeoning hippie scene in the Haight. He moved to Chicago for two years of college at Northwestern, then enrolled in the San Francisco Art Institute. He described the years that followed not in terms of his twenties or thirties but by obsessions. First it was sculpture; then he took up rhythm guitar and punk music, headlining a band with a suitably gearhead name, The Pliers. His best song, he said, was "Baby Don't Mistake a Hard-On For Love."

Then Felk fixated on acting, intending to get the experience for voiceover work. He took acting classes at City College and joined local productions of the plays *Bleacher Bums* and Steven Sondheim's *Assassins,* and brought that same intensity to establishing himself onstage.

But mechanical obsessives tend to abhor working in large groups, where their control over a project gets diluted. Felk was no exception; he gave up on acting and concentrated on his profession, which was carpentry, and thought about playing music again, and generally continued his search to recapture the feeling of competition and belonging from his Soap Box Derby days.

In August 1996, he was driving home from a carpentry job in San Francisco's Sea Cliff neighborhood, navigating the city's expensive Marina district, when he saw a sign for Robot Wars.

To Stephen Felk, who was led by his gearhead curiosity to park his truck and wander inside that Friday night, Robot Wars was a revelation. Here were dozens of other mechanical obsessives, all plying their trade, but also collaborating and competing. "It was really obvious, seeing all those guys, that there was something out of the ordinary going on there," he said. "It wasn't like at an auto race where everyone is secretive about their cars. Right from the start you had this great feeling of camaraderie, where everyone would love to explain all the stuff about their machines."

He watched that night, then returned Sunday for the finals. The event was sold out. A guard standing watch outside apologized, but Felk kept bugging him—he had to see it, he had to be let in, it was too important to miss—until the guard couldn't take it anymore.

"Look, why don't you just go down the side of the building," the guard said. "I'm going to turn my back, do whatever you want."

Felk, who had never snuck in anywhere in his life, slipped into the side door of the pits, pretending he was a builder.

Which, of course, he immediately thought of becoming.

After the event, he couldn't sleep and took long walks, thinking through all the possibilities for designs and weapons. At the same time, he knew it was dangerous, that he could literally lose himself and great chunks of time and money to this project.

As he was walking, he found a 1941 nickel on the pier at Fort Mason where Robot Wars had been held days earlier. His dad had sailed from that very base to Guadalcanal in 1942. His father was 41 when he died; he might well have sailed with the nickel in his pocket. It was a sign. "I just knew it," Felk said. "I know it sounds weird and wacky and all that, but when it happened, I just knew it was a communication from my father saying, 'Do this.'"

Felk returned to the one-bedroom apartment in Nob Hill and started working on his own unconquerable dream robot.

Building a radio-controlled combat robot was not as easy as first appeared. There were many complicated choices: What materials do you use? Where do you get your motors? How much do you spend? How do you ensure that the robot fights with the appropriate combination of speed and power—that it doesn't move too quickly with too little control, or have adequate push but be too sluggish and easily outmaneuvered? Should he use tank-style steering, where all four wheels cause the machine to turn and give the driver greater control, or automotive steering, where only the front two wheels turn, which allows the machine to drive in a nice straight line but turn clumsily.

Felk started thinking about all of this, and he devoted himself completely to the problem, the challenge of building a fighting robot. For months there was nothing else in his life—just the day job, carpentry, and then at night, the robot.

He sketched out several designs and settled upon combining the wedge shape of the successful La Machine with a lifting arm like that of the champion Biohazard. But there'd be other features too: If his bot got flipped over, the shields that protected the wheels on the top of the robot would rotate open, allowing Felk to drive his bot upside down. On the very end of the lifting arm, teeth would articulate upward, allowing his bot to grab the underside of enemies and drag them around the arena. That was the plan: He would build the whole thing in one year, for Robot Wars '97, by himself, with no compromises.

Like many gearheads, Felk was superstitious. He believed in signs, and all of them were encouraging, even eerily so. One of the first calls he made before beginning construction was to a wheelchair store in the city's Richmond district, and the vendor just happened to have a Hoveround wheelchair in stock that he couldn't fix and was willing to sell for a hundred bucks. It was a huge discovery. Once Felk bought it and took it apart, he saw that it could supply the components for half of the drive train and the entire lifting assembly.

He didn't have a computer. He couldn't sketch out his idea on CAD design software and contemplate it in three dimensions on a screen. So he spent months building test models, called "mules," to try to figure out the proper power-to-speed ratio. And he watched videos of prior Robot Wars events, the ones edited by Trey Roski (which predictably luxuriated in La Machine's victories), and he weighed how to beat the various veterans. He bought a toy R/C car to practice driving and a drill press for his kitchen, and started cutting up big pieces of aluminum.

Nineteen-ninety-six turned to 1997, and the next Robot Wars was eight months away. Every night there were assorted design decisions, a whole series of variables to weigh. It was going to be a two-wheeled bot, so what would keep the nose from dragging on the floor? Felk had to design miniature wheels, called casters, for the robot's beak.

Then it was suddenly summer, not that you could tell in fog-locked San Francisco, but still, the event was two months away. After a while, Felk stopped taking carpentry jobs and started working exclusively on Voltarc—he pinched the cool name from the neon lamps in the elevator of a San Francisco building he was working in. He was going to do it right, his unconquerable dream robot.

Most mornings, he woke at eight o'clock and jotted a list of six things to accomplish that day. He would start with the first thing (say, milling the teeth for his lifting arm), and by the time he finished it would be one o'clock in the afternoon. Then he would get going on the second thing (attaching the casters), and suddenly it would be six at night. He'd take a walk and get a sandwich and get back to work at eight; he was going to execute the entire vision. He would tackle the third item on the list and suddenly—how did the day slip away?—Jay Leno would be doing his monologue. Then it would be three in the morning, a 19-hour day, with no contact with other human beings and five hours of sleep until the next day, and a new list of three old tasks, and three new ones.

Felk started to realize that these combat robots really did kick ass—their maker's.

Then there was a week left before the event, and he mailed in his $75 application fee, and then 72 hours left, and then 48, with six new things on the list every day and only three done after 19 hours. He was losing his mind. Those five hours of regular sleep diminished and dwindled until he was hardly sleeping at all, going 100 miles per hour trying to finish the robot, chasing the perfectionist's shadow—an enticing vision of how things could be but aren't, so get back to work . . .

. . . until the day before the event, and Felk looked at Voltarc, which he had constructed methodically over the course of a full year, never compromising, and what he saw was not the unconquerable dream robot but a half-finished shell of aluminum. The drive train was running but the control system hadn't been wired up. He hadn't tested the robot. The casters were in place and those cool rotating wheel shields worked, but they had not been tested under battle conditions (and in fact would later require a complete rebuild). He had worked nonstop for a year, and Voltarc was an unfinished hodgepodge of motors and metal.

One day before the event, he took a deep breath and admitted to himself that he wouldn't finish in time. He had drastically underestimated the amount of work building a dream robot entailed. "I didn't want to compete with just half of Voltarc," he said. "I wanted everything to be working. I thought it was going to be a really cool machine."

He slept for the next day, than dragged himself to Fort Mason. He found Marc Thorpe, who looked exhausted himself, and told him he hadn't finished and couldn't compete. Thorpe gave him tickets to watch. For the second year in a row, Felk sat among the spectators instead of walking amid the builders. This time, he watched in agonized frustration and self-loathing, and it was nothing like being 13 years old and racing Soap Box Derby cars.

After the event, he took a short break from robot building, then plowed back in, eyeing the 1998 competition. But that event, and the whole Robot Wars tradition, was about to be consumed in legal turmoil.

To Felk, watching the courtroom drama unfold would prove just as devastating as his own '97 debacle. He wanted to feel a part of the robot combat enterprise, to meet all those great fellow obsessives, and most of all, to test his gearhead mettle. *He didn't even know how Voltarc would do in battle, and now they were talking about shutting it down?* Like all the other robot builders, he began to realize what a big part of his life Thorpe's idea had become.

Their desire to compete constituted a collective will: Whatever the circumstances, the gearheads would not let robotic combat disappear.

In fall 1997, Marc Thorpe dealt with the consequences of his decision in the jury room of a New York federal court. Instead of walking away from Steve Plotnicki, he had paid one dollar for a license that allowed him to put on Robot Wars '97. Now his future, and that of the community he created, was as inextricably linked to Profile Records as ever, and he was being asked to live up to his end of the bargain: The deal in New York had only been a precursor to finally concluding negotiations and turning the joint venture into the more permanent Robot Wars LLC.

But once again, Thorpe couldn't take the final step. He looked at the new papers drawn up by Profile's lawyers and felt that bits of what he had agreed to in Judge Batts's courtroom had been subtly and scurrilously altered. He worried most about a provision that would allow Profile to foreclose on the Robot Wars assets if the new LLC didn't pay back, on a specified schedule, the more than $300,000 Profile had loaned the partnership over three years. Thorpe feared the agreement might give Plotnicki an incentive to sabotage the business, so he could take the whole thing over. Thorpe simply didn't trust his partner.

The parties met in two days of mediation in December 1997, and a new agreement resulted. Thorpe would sell his interest to Profile for $250,000, get a 10 percent royalty and a license to put on two events each year in the United States.

But then the California artist had another crisis of confidence. "When the documentation arrived, it was at substantial odds with the terms I believed had been reached with the mediator," he claimed in court documents. In the new documents, the royalties were limited to just toys and games—not all merchandise, which he thought he had agreed to. The noncompetition clause also seemed overly broad, and his license to hold an event was nonexclusive. It was the same old fear: Profile might sell the idea to someone else, who could then stage an event in the same city on the same weekend as Thorpe's event and push Thorpe out altogether.

Again, Thorpe refused to sign, and he implored Profile to let him put on the '98 Robot Wars event anyway.

"As you know from the recent financial statements we sent you," Plotnicki wrote him, "the venture is currently insolvent and does not have the money to pay for the staging of the 1998 event."

When his lawyers asked Judge Batts to force Profile to issue Thorpe another license, so he could hold the event that August, Batts refused. She had been on this merry-go-round before and declined to stretch the joint venture agreement, which called for both parties to agree on all business decisions. "Apparently," she wrote, "there is no forum currently known to the modern judicial system which can assist these parties in signing a final agreement."

That spring, Thorpe appealed his case to a three-judge panel in New York, hoping to get permission to put on Robot Wars '98. It denied him as well. Time was running out.

At another settlement meeting that spring, Thorpe offered to buy Profile's share of the business for $500,000. This brought howls of incredulity from Plotnicki. The New York music exec had premised his tactics on the fact that Marc Thorpe was a man of extremely limited resources, who would be unable to endure an extended period of litigation. It was another version of jungle law: A man like Thorpe had no business standing up to a man like Plotnicki. Now Thorpe wanted to buy Profile out? "With whose money?" Plotnicki demanded to know.

Thorpe had found a sympathetic ear for his troubles in Trey Roski. The Robot Wars creator approached the driver of La Machine that fall, after the fourth Robot Wars competition. Thorpe's previous attorney, paid for by Bob Leppo, was leaving the case, and he had no money to hire a new one. Could Roski help?

Not only was Roski in a unique position to help, he also possessed an insouciance toward money that usually comes from having plenty of it. "I just call up Majestic Realty, ask for $10,000, and it mysteriously appears," Roski liked to say. "I don't even know how much I have."

Roski was only too happy to help.

In October 1997, Roski gathered Thorpe, his lawyer, and venture capitalist Bob Leppo at his dad's office in L.A.. His father's attorneys, from the high-powered law firm of Latham and Watkins, were present and listened to Thorpe's story.

The California artist told them he was near the end of his rope—physically, emotionally, and financially. He had been living off loans from credit cards and had refinanced his home to pay down the expensive consumer debt, which quickly ballooned back up. He was taking medication four times a day, with the tremble in his right side worsening and his stamina declining. He didn't want to be in business with Steve Plotnicki any longer.

What could he do? He couldn't go out and get another job in the special-effects industry. Shops like ILM were now trafficking largely in computer graphics, while the old animatronics shops were being downsized. With his limited dexterity, he couldn't compete with the younger model makers. And the original joint venture agreement, and his subsequent agreement in Judge Batts's court, had him locked in a room with Plotnicki. He couldn't just go out and start another competition.

Ed Roski, Jr., came to that meeting late and left early. Afterward, the business tycoon told his son, "Nobody gets involved in someone else's lawsuit." But the younger Roski wanted to help; he desperately loved the sport, and wanted to be involved in every way possible. Against his dad's wishes, he would provide financial backing and personal counsel to Thorpe over the next year. They spoke on the phone constantly.

That fall, Thorpe spun various schemes before his new benefactor, plans that he thought could benefit them both and end the deadlock. Thorpe talked about joining forces with Roski, becoming partners in a new business whose primary asset would be the 50 percent ownership share of the Robot Wars joint venture. The new business would get licenses from Profile to hold events.

To Steve Plotnicki, conditioned by the pattern of his past disputes, the whole Robot Wars saga was starting to smell like the Run-D.M.C. and Cory Robbins scenarios all over again. Thorpe did not have the financial wherewithal to be fighting the battle for control by himself. Someone else must be standing behind the scenes, helping him. Suspicious, Plotnicki sent his lawyers to depose Roski and Bob Leppo in spring 1998.

"I just wanted to compete," Roski said in his deposition. "I'm in second place right now, so I have vengeance to get back at someone, and now they are not going to let me play." Later in the deposition he said, "It seemed like Marc needed help. It was upsetting to think that somebody like this was taking advantage of a friend."

Roski and his father's lawyers from Latham and Watkins spent hours across days and weeks going over the remaining alternatives with Thorpe, trying to find a way out of the stalemate. They couldn't buy Profile out, Thorpe refused to renegotiate the joint venture agreement, and he didn't want to give up control of the competition. Meanwhile, the case had returned to Judge Batts, who Thorpe feared might find that he had acted in bad faith by refusing to sign the tentative '97 settlement. What was left?

Well, said the Roskis' high-priced attorneys, the American legal system does provide an option for citizens trapped in a downward financial spiral. It's called Chapter 11 bankruptcy—and according to federal law, it provides the added benefit of stopping all other litigation while allowing a debtor to dissolve detrimental contracts.

At Bob Leppo's urging, Marc Thorpe e-mailed the robot warriors that March. The venture capitalist thought Thorpe owed it to the community to let them know that Robot Wars '98 might be canceled. In the message, which infuriated Steve Plotnicki, Thorpe tried to explain why there had been no contact from tournament organizers since the last competition:

> We are still trying to come to an agreement, but Profile has refused to allow me to move forward with the event unless the final documents are signed. That arrangement continues to be unacceptable to me due to the number of unresolved issues and the time required to resolve them.... And, besides, no one benefits by the cancellation of the event in any case. I have worked far too hard and too long and at far too much personal sacrifice to be bullied into submission by a wealthy businessman and his clever attorneys. Also, I am not one to succumb to tactics of coercion and intimidation. Thankfully, few have suggested that I do. And, fortunately, a few individuals have helped me in ways that simple gratitude cannot address.
>
> ... I believe that [Plotnicki's] strategy is as follows: He knows that the bigger Robot Wars gets the higher the value becomes and the more difficult it will be for him to get control. He might not ever get control if Robot Wars were financially prosperous. So lately he has been doing all he can to avoid deals and opportunities and is generally withholding approval for everything including Robot Wars '98. He has some hollow excuses but the reality is as obvious as an elephant in a kitchen.

Thorpe ended by urging competitors to write Judge Batts with their feelings on the potential cancellation of another year. He signed off, "Keep the faith! You build . . . I'll fight."

Steve Plotnicki would later see that e-mail as the beginning of another conspiracy. He thought he recognized more at work than just a tenacious, frustrated California artist sending an e-mail to the community he created. Plotnicki felt that the message constituted a deliberate attempt to ruin Profile's standing among the robot makers. To him, the e-mail would prove itself a kind of temporal D-Day, spreading destruction years ahead into the future.

"What happened with Gary Cline, Carlo Bertocchini, and everything else after that was just derivative of that e-mail," Plotnicki said. "That letter pretty much encourages the inmates to take over the asylum, which is exactly what happened."

One month after Marc Thorpe sent the e-mail, Carlo Bertocchini was surfing the Web from his home in Belmont when he came across the free community bulletin boards of the Delphi online service. He signed up on a whim and dubbed his board "the Robot Wars forum." It took him about 10 minutes to enroll, and would cost him months of trouble and thousands of dollars afterward. He intended the forum to be used for an exchange of technical and strategic information. Within weeks, there were 200 messages under the headline, "How to beat Blendo."

But the robot builders also began to swap information on the lawsuit, and to discuss the possibility that Robot Wars '98 might be canceled. "It became a place to find out what was happening and how to plan for the next event, if there even was one," Bertocchini said. By June, 5,000 messages had been posted.

The builders were no longer just seeing and speaking to one another once a year. With the help of the Internet, they were now a community. Veterans like Bertocchini, Dan Danknick, Supreme Commander Tilford, Jim Smentkowski, and many others began trading messages and speculating about the sport's fate. Newcomers like Stephen Felk (who had to borrow a computer) monitored the boards to keep up with the fast-developing legal situation.

There were others watching the Robot Wars forum as well—the attorneys of Profile Records. One month after the bulletin board was created, Steve Plotnicki's attorney, Fran Jacobs, sent Delphi a letter asking that no further use be made of the company's trademark. Delphi's lawyers contacted Bertocchini, who agreed to acknowledge on the site that the trademark was registered to the Robot Wars venture. But the maker of Biohazard made it clear that he didn't feel obligated to change the name, "Any more than the evening news is obligated to state that they don't own the NFL," he wrote back to a Delphi exec. "The business of Delphi depends on the ability of each of us to exercise our First Amendment rights. I am sure your attorney will not want to set a precedent of kowtowing to any yahoo that comes along and threatens to sue."

This was passed along to Fran Jacobs, who responded by insisting the name of the bulletin board be changed to "An unofficial forum on Robot Wars events."

Bertocchini responded, "I know of no legal precedent that requires the owner of a web page to place certain words in certain prescribed locations on their page. Our disclaimer is clearly visible, and as such, I think it satisfies any requirements for the disclosure of the nature of the forum."

Steve Plotnicki called Thorpe that spring and tried to get him to pressure Bertocchini to get a license to operate the online forum. Annoyed, Thorpe told Plotnicki, "Carlo can do whatever he wants."

Meanwhile, heeding Thorpe's missive, the robot makers responded with another round of supportive e-mails to Thorpe and to Judge Batts in New York City.

"If your 'partner' is able to get control of Robot Wars, it will never be that vision you intended it to be," wrote Christian Carlberg, a new competitor at the '97 event with the walking lightweight Pretty Hate Machine. "Instead of a fun event for builders and spectators, it will become another corporate bottom line."

"I wonder if Profile Records realizes that the participants in these events are a close-knit group?" Todd Mendenhall, an L.A.-based aerospace engineer, wrote the judge. "I believe they will support Mr. Thorpe and would not enter a Robot Wars event if he is not associated with it. The bottom line is that even if Profile manages to obtain full control of the event, they will find that they have no contestants."

Robot maker Adam Clark wrote to Thorpe from the United Kingdom, where, after two years of development, the British television version of Robot Wars was finally starting its run. "I'm sad to hear about all the dealings going on with the U.S. competition this year. It's a vision and an idea that gives too many people such good times that it deserves to stay around a long time." Clark noted that the following day, he was having 30 friends over to his house to watch videos of the U.S. competitions. He ended his letter by quoting a scene from *Jurassic Park:*

Henry Wu: "You are saying that a group of animals, entirely composed of females, will breed?"

Ian Malcolm: "No, I am merely stating that uhh . . . life finds a way."

"Profile might stop the event," Clark wrote, "but they can't stop the combat."

Steve Plotnicki had not been born yesterday. Monitoring Carlo Bertocchini's Robot Wars forum from New York City, he suspected that the robot-building community might try to stage an alternative competition in place of the deadlocked Robot Wars. So he too joined the fray, writing the builders two e-mails that spring, which were posted to the Robot Wars forum. "I've tried to refrain from writing to you because, despite Marc Thorpe's attempt to make our dispute public, it's a private, commercial dispute that can't be solved by anyone but the two of us," Plotnicki wrote. He too reviewed the history of the business, but from his perspective, Robot Wars had been a financial burden. "Money was a one-way street; all of it out, all of it ours." Nevertheless, he wrote:

> We foresaw the day where we could turn Biohazard, La Machine, or The Master, as well as other robots, into house-

> hold names like the "Ninja Turtles" or other great action characters. Secondly, Marc promised to give us additional shares in the company for the money that we put in beyond our original investment. As part of our agreement, Marc would continue to hold the San Francisco event. We thought all the issues were resolved, but then came the messy part.
>
> Marc refused to sign documents that reflected what he had agreed to. Even though Marc had received the benefit of using our money, including putting nearly $100,000 in his own pocket, it turned out that all along he was leading us on, making promises that he wasn't going to keep. To further aggravate things, he arranged to hold Robot Wars '97 behind our backs. . . . We were left with little choice but to bring litigation against him.

Plotnicki concluded with a veiled threat toward anyone who might try to hold another competition: "At this point, the only way Robot Wars can be held is if Marc and I come to terms. . . . We believe that the court will see through any attempt to violate its order, and will take the appropriate actions. If that happens, it would be unfortunate for an innocent third party to be caught in the middle. Marc has known for a long time that we would not consent to hold another Robot Wars event until we had reached an agreement."

Gary Cline read this but didn't heed it. Cline, the Marin, California–based maker of the simian-armed robots Spunkey Monkey and Mad Monkey, thought he could help out. He was a 51-year-old computer programmer and had two grown sons and a deep and intractable love for the new sport.

He called Marc Thorpe. "What if somebody else put on an event?"

Thorpe replied, "Well, I'd have mixed feelings about that. But one thing I know for sure, I couldn't be involved in it, directly or even indirectly."

Cline didn't call back until later that summer, after the Second Circuit Court of Appeals in New York denied Thorpe's appeal to hold Robot Wars '98. Cline told him, "I'm going to do something. I just want you to know, I mean well."

Thorpe thought Cline was going to file a class-action lawsuit against Profile Records, on behalf of the robot builders. Instead, Cline posted a message on the web forum, unveiling his own event, Robotica.

Robotica, Cline said, was going to be a party—no paying spectators, just builders and their robots. It would be a short-term solution to the cancellation of Robot Wars. An *L.A. Times* reporter called Cline at his house in Marin and wrote a feature story on the new event. "The thrill is hard to describe," Cline told the paper. "It's better than fishing, a whole lot better than baseball. It's fighting and it's not fighting. It's just boys at play."

Robotica was intended to be staged by the competitors, for the enjoyment of the competitors. Cline hoped to sell T-shirts and videotapes to defray some of the costs. He gathered his robot buddies to help him prepare for the competition. Three weeks before the event, Dan Danknick picked up 1,200 pounds of polycarbonate plastic from a supplier in Orange County and drove 200 miles to Cline's house. Midway, he pulled off the road and slept in his truck. The next day, they had a building party at Cline's house, hammering together an arena made of plastic and plywood. Jim Smentkowski, the builder of Hercules, was there, as were Trey Roski and Greg Munson of La Machine, and Patrick Campbell of Frenzy, and a new competitor named Ray Scully. Cline's wife, Laurie, served pizzas, chips, and dip. They built the parts for the arena and stored it.

Cline spent $15,000 of his own cash preparing for Robotica—he was that devoted to Thorpe's idea. He had rented out the Cow Palace, a six-acre amphitheater in South San Francisco. Built in 1941, it was initially used as a launching station for troops embarking to the Pacific Theater during the war. The Beatles later performed there and now it would resuscitate the new sport of the next century, robotic combat.

But in retrospect, Cline—another mechanically inclined, legally naïve gearhead—did a few things wrong. On his website, a "guess-who's-coming-to-Robotica" list began with Marc Thorpe and also included Trey Roski. A copy of the rules on the site appeared nearly identical to the Robot Wars rules. He also posted freely on the Robot Wars forum, appealing to builders to attend his event, enticing them with the prospect that Blendo, the feared and indestructible spinner, would compete. Cline talked freely on the forum about his plans to market a video from the event to recoup some of the costs and perhaps even make some money.

Cline's plans were starting to interfere directly with Steve Plotnicki's strategy for breaking the deadlock. Plotnicki was talking to the U.K.-based Mentorn Films about coming to San Francisco and staging a replacement Robot Wars event, without Thorpe's involvement. The event would air as one episode of the new Robot Wars game show on BBC 2, which had just debuted to an audience of nearly four million Brits. On the Delphi forum, Mentorn's chief, Tom Gutteridge, floated the possibility of coming to the United States, and the American builders reacted with venom. "Who is going to sell their soul to be on the BBC?" wrote Derek Young, maker of featherweight Mr. Smashy, in one typical post. "Forget that. Robot Wars is dead. Long live Robotica." Gutteridge rescinded the proposal.

It was all too likely that a legally minded Robot Wars shareholder with an obsessive desire to defend his trademarks could look at Robotica and think, "This is a lamb named Robot Wars dressed in the clothing of a sheep called Robotica."

That summer, with preparations for Robotica in full swing, Steve Plotnicki called Gary Cline at his home in Marin to discuss the matter. His goal: to get Cline to admit Thorpe knew about this alternative event, since Thorpe was legally bound to divulge competing entities to Plotnicki under the joint venture agreement.

Cline and Plotnicki had a short, unpleasant conversation, in which Cline admitted he had spoken to Thorpe but insisted that the artist wasn't involved in Robotica. Plotnicki kept pressing him for more information. The call ended when Plotnicki told Cline he would get what he wanted in the courts if necessary.

Plotnicki, sensing a conspiracy, believed Cline was putting on "a de facto Robot Wars on Marc Thorpe's behalf." Cline told Plotnicki to "fuck off" and slammed down the phone.

One week before the event, at 5:30 A.M., there was loud banging on Gary Cline's door. "I know you're in there!" someone yelled from the front driveway. Cline's wife, Laurie, was home alone that morning. The court server hurled the court papers at the door and scared the bejeesus out of her. "She was really distraught," Cline said. Later, he would rage to the *San Francisco Weekly*'s Jack Boulware: "It was a party. That's all. The fucking asshole sued me for [millions]. It's like, forget it, you know? He blames everybody else for his bad reputation."

Up in Novato, Trey Roski was also named in the lawsuit, and charged with conspiring to undermine the value of the Robot Wars trademark. However, he was not served with the complaint, and wouldn't have to fight—yet.

Fifty miles south in Belmont, California, Carlo Bertocchini was charged with violating the Robot Wars trademark on the Internet. The maker of Biohazard was served with the complaint and chose to fight. Like Thorpe, he also found a willing patron in Trey Roski, who loaned him $10,000 to hire an attorney. Bertocchini also raised $3,500 through a collection from the other builders over the Web forum. With those funds, Bertocchini hired a New York lawyer and respond to the charges.

Gary Cline, already licking his wounds, chose not to fight at all. "I didn't want to throw good money after bad, so I just curled up my tail and whimpered away," he said.

One week before the event, the Robotica website read, "Due to legal reasons, plus overwhelming financial burden, the Robotica '98 party is hereby canceled."

◘

The builders came to San Francisco anyway. They had spent the money on hotel rooms and plane tickets for Robotica, so why not? Ilya Polyakov came from New York, Derek Young from Vancouver, Mike Winter from Wisconsin. They came even though there was really nothing to come to. Jason Bartis drove up from Los Angeles; he'd already spent more than $3,000 on his latest robot, plus hundreds of hours, and had convinced family and friends to fly in from around the country. He was now the owner of about two dozen T-shirts that said "Robotica '98" on them. Everyone brought their robots.

If there was any doubt about who their common enemy was, the latest round of lawsuits had cleared it up. Hearing Thorpe complain was one thing, but learning from Cline and Bertocchini that they had been sued—and then having Robotica canceled—represented an assault on the community itself.

Plotnicki was now a villain of the highest order. "If anyone from Profile is reading this," Alex Rose of Rhino Halon wrote on the Web forum, "I hope you realize how dangerous a set of enemies you just made. The New York mafia will look tame."

The gearheads came to San Francisco, more than 100 in all, and they gathered in an empty dance studio in the South of Market neighborhood. They were all there, veterans like Bertocchini, Gage Cauchois, Mark Setrakian, and Trey Roski, and new builders like Stephen Felk, who attended with his finally completed Voltarc. Many hauled their robots up the narrow stairway to a room on the second floor. "For a lot of us, just looking at the robots and talking to the other builders is more than half the fun anyway," said Todd Mendenhall, the meeting's organizer.

They were meeting under the auspices of a union of sorts, a rules body formed from the discussion on the Web forum. It was called the Society of Robotic Combat, or SORC. Its purpose was to serve as a collective voice for the builders in these turbulent times. Mendenhall, who had not yet competed in Robot Wars, offered himself as president of the new organization and was elected through an e-mail vote. Part of his job was to explore putting on an event on behalf of the builders.

Monitoring the forum from New York, Steve Plotnicki had learned about SORC and called Mendenhall the week before the gathering in San Francisco. Would SORC care to discuss the possibility of buying a license from Profile to put on an event, while the Robot Wars partnership was still hung up in court? Plotnicki thought this might mollify the builders desperate to compete and give him more time to settle with Thorpe. Mendenhall told Plotnicki he would bring it up at the meeting.

Mendenhall opened the SORC meeting gingerly—with a show and tell. Builders stood up, dragged their robots to the front of the room and discussed their technologies and features. Christian Carlberg from L.A. talked about his new robots, Little Slice and Nasty. Eleven-year-old Lisa Winter, daughter of Mike Winter, demonstrated Doughboy, a square box with Pillsbury doughboy mitts covering the lawn-mower blades on top. Stephen Felk talked about the doggedly constructed Voltarc.

After more than an hour, the show and tell ended and Mendenhall raised Plotnicki's proposal. SORC could put on an event—if it got a license from Profile. "What kind of terms is he talking about?" people wanted to know. "We didn't talk about terms," Mendenhall answered.

Then it was Marc Thorpe's turn to speak. Thorpe too had read about the builders meeting on the Web forum and called Mendenhall to ask if he could come. He wanted to explain his side of the story. He couldn't say much about the state of the legal dispute, but he asked the builders to have faith that it would soon be resolved. He explained that he had just filed for Chapter 11 bankruptcy and that he had a good chance of getting a judge to sever his contractual ties to Profile Records.

"Marc appeared very beaten down," said Mendenhall. "It looked like he wanted to be in a cave."

After Thorpe finished talking, the issue was thrown open for debate. Should the robot community deal directly with Plotnicki, without their besieged leader? Veterans like Carlo Bertocchini and Mark Setrakian stood up and vowed never to participate in a Profile-sponsored event. Jim Smentkowski, the maker of Hercules, warned, "Any deal with the devil is a still a deal with the devil." A few others didn't care and wanted to fight robots at any cost.

Trey Roski was silent the whole time.

The issue was brought to a show of hands. The builders overwhelmingly voted that SORC should not seek a license from Profile for any event. "Some of the members voiced some opinions about what Profile could do with their proposal, which I can't repeat," recalled Carlo Bertocchini.

The rest of the meeting was informal. Builders walked around, talked to each other, and shared tips about their robots. After more than three hours in the studio, builders filtered outside onto the street. A few drove their robots up and down the sidewalk and marveled at Oakland programmer Jonathan Ridder's new robot, Ziggo. Named for Ridder's maladjusted and since-deceased cat, Ziggo was essentially a lightweight version of Blendo, the furiously spinning wok.

The next day, Mike Winter and his daughter Lisa met Mike's two brothers Rik and Steven at a parking lot in San Leandro. Will Wright and his daughter Cassidy were there too. They rented a van, bought plywood and wire mesh for an arena, and made two trophies out of discarded metal objects they found diving in Silicon Valley trash bins. They brought a portable electric generator and a boom box.

The event was top secret. Only a few lightweights were invited. "We didn't really tell anybody about it," Mike Winter said. "At that point, it seemed too legally dangerous. It just was unimaginable to us that the robots would not fight each other."

Profile might stop the event, but they can't stop the combat.

Six robots were there, including the newcomer Ziggo. Like Blendo, Ziggo destroyed the competition. He tore into Winter's bladed walker BORB and turned it into little pieces. They also played a game of robotic hockey, and Ziggo (an excellent goalie) won that event too. There were two trophies and Jonathan Ridder collected them both. About 15 spectators wandered by and the builders made them sign release forms and wear face shields.

Across the Bay, Trey Roski and Greg Munson had the same idea. A few months after the SORC meeting, with the Robot Wars business still frozen in the courts, the cousins headed out to a highway overpass in Novato. They called their event the "Underground Robot Street Fight" and they got the local news program *Evening Magazine* to tag along and cover it. Roski and Munson brought La Machine, plus a new robot they had spent the last year building, Ginsu, which had a clear Lexan frame, framed by aluminum, and four saw blades for wheels.

Scott LaValley was there with his robot, DooAll. Jim Smentkowski brought his new bot, Junior. Lowell Nelson and his father, Steve, former commercial fishermen from Quincy, California, drove

250 miles through the rain. Stephen Felk came as well and, for the first time, got to fight with Voltarc. At one point that afternoon, Voltarc got under La Machine and hoisted it up into the air with that specialized lifting arm; Jim Smentkowski took a picture of it and put it on his new website, robotcombat.com, with the caption, "Voltarc teaches La Machine a thing or two." Stephen Felk loved that. It almost made two and a half years of obsession and disappointment worth it.

It was all for fun, and it was a small mutinous act against Profile Records. Later that day, Trey Roski told the gathered builders about another mutiny, this little idea he was working on.

Chapter 6

Thorpus Delicti

I know many of you are depressed and discouraged by what has been happening as of late. Don't lose faith. I am sure, whether it is with Marc Thorpe or Profile or Mentorn or SORC, something WILL happen. It may happen in San Francisco, Los Angeles, New York, UK, Las Vegas, a hobby shop parking lot or even in your driveway. It may happen tomorrow, in a month, in a year, or even in three years. The passion for this sport burns too brightly for its light to flicker out.

—From "Zen and the Art of Robot Wars," posted by Peter Abrahamson, builder of Gigan II, in a message to the U.S. robotic combat community on January 24, 1999

"Why did the robot cross the road?"

"Why?"

"She was programmed to do so."

Groan.

"Why was the robot attracted to Margie?"

"Why?"

"Because she was a magnet."

Ugh.

It was October 1998, two months after the canceled Robotica event and the subsequent SORC meeting at the South of Market dance studio. A dozen robot builders stood in a parking lot in Santa Rosa, telling bad robot jokes. They had traveled from across California to a courthouse 60 miles north of San Francisco. Dan Danknick,

Peter Abrahamson, and Christian Carlberg drove up from Los Angeles; Jim Smentkowski, Gary Cline, Stephen Felk, and others came up from San Francisco. At 9:00 A.M., they headed in to watch—they hoped—Trey Roski buy Robot Wars and end the Ice Age.

The previous May, Marc Thorpe had filed for bankruptcy in the Santa Rosa division of U.S. federal bankruptcy court. He was looking for an escape route: Steve Plotnicki had sued him in New York federal court, refusing to allow him to hold any more events until he renegotiated their partnership, and Judge Batts and the federal appeals court had ultimately sided with the music exec. Thorpe was essentially trapped. He could fight on in New York, but he wasn't in financial or physical shape to be litigating on the other coast anyway. Or, he could give up, renegotiate along the terms Plotnicki wanted, and lose control over his revered creation.

Bankruptcy was a third option. In bankruptcy court, Thorpe hoped to reorganize his family's loans and bills—and to convince a new judge to nullify his contractual relationship with Profile Records, so he could restart the Robot Wars business afresh with Trey Roski. "It became clear that it was impossible to deal with them," Thorpe said later in a deposition. "Legal costs and everything were getting out of hand. That, in addition to my bank cards, to the personal loans, to the lack of income as a result of having to tend to all of this, and my health issues . . . our situation became untenable."

Dipping back into the family well, Roski loaned Thorpe $150,000 to hire a prominent San Francisco bankruptcy attorney, William Weintraub.

Judge Alan Jaroslovsky, a Navy veteran of the Vietnam War who was in the tenth year on the federal bankruptcy court bench, got the case. When the judge refused to dissolve the contract outright, Weintraub proposed a common plan for bankrupt estates with a valuable asset. The plan asked for the judge to dissolve the Thorpe-Profile partnership and to auction the Robot Wars trademark to the highest bidder. The auction would work like this: Trey Roski, serving in the role that bankruptcy law calls "the stalking horse," would bid first for the trademark, at $250,000. Anyone could offer a competing bid, but Roski could choose to outbid any party who matched him. Profile immediately filed a motion objecting to the auction.

The robot makers congregated in Santa Rosa to watch the auction. They had read about the impending court date on the Web forum, and on the Robot Wars website, which Trey Roski and his design company, Impact Media, still controlled. Trey was there with his father, Ed, and their powerful lawyers from Latham and Watkins. Thorpe was there with his wife, Dennie, and attorney William Weintraub. Profile was represented by Marin County attorney Bill Pascoe. It was supposed to be one final, climactic confrontation of all the feuding parties.

But it didn't happen. Jaroslovsky said he needed more time to review the case and set a new hearing date for the following month. The builders had congregated in Santa Rosa for nothing. They went for a late breakfast instead.

One month later, on the new court date, several builders again showed up to support Thorpe and Roski. Again, Trey Roski sat with his father, and again the judge wouldn't approve the auction plan. Instead, Jaroslovsky met with the parties and advised Thorpe's counsel that the joint venture agreement was binding, and that he couldn't allow the auction to take place. He urged Thorpe to settle with Profile. Thorpe, Plotnicki, the Roskis, and their lawyers spent the rest of the day searching for another way out.

It didn't look like it was going to be resolved any time soon, so Marc Thorpe tried another approach. He heard from a former ILM colleague about an experienced New Hampshire attorney named Frederick Fierst, who specialized in licensing matters and had represented the creators of the Teenage Mutant Ninja Turtles. Thorpe went to Roski and got him to fund one last attempt at saving Robot Wars in the United States. Fierst's mandate was to work out a deal between Thorpe and Plotnicki, and to get a license so that Thorpe and Roski could go into business together, holding annual robotic combat events in the United States.

Frederick Fierst spent the last two months of 1998 talking to the entrenched sides. He spoke to Thorpe and found the California artist worried about the effects of stress on his disease and doubtful the music exec would be willing to treat him as an equal partner under any settlement. "Marc was obsessed with this Robot Wars property that he had created and nurtured," Fierst said later in a deposition.

"He was really, really particular about what was said or done or put into any agreements to the *nth* degree. It was extremely difficult to bring Mr. Thorpe along to a resolution."

Fierst also talked to Plotnicki and found the New York music exec insistent that Marc Thorpe should have to rehabilitate his reputation in the eyes of the robot makers. "Marc has badmouthed me and done damage to me. And I need to have his commitment on some certain level to rehab me in the eyes of the robot makers in order to make this whole business work," Fierst recalled Plotnicki saying.

Plotnicki was also extremely wary of the Roski family. Their lawyers, he pointed out, helped the Roskis and partner Philip Anschutz buy the L.A. Kings hockey team out of bankruptcy. Perhaps they were trying to do the same thing here, he reasoned. "They're going to have to first pay me back 100 cents on the dollar on all the money I've lost fighting with Thorpe. And I'm going to sue them when this is all over, and I'm going to collect every single penny of my damages from them, as I always do from everybody. And only if they're prepared to step up to the plate and settle up . . . do they have any hope of getting a license" to put on a robotic combat event.

Plotnicki also refused to suspend the leftover claims against Roski from the 1998 lawsuit. Getting the message, the family dropped out of the negotiations. If Trey wanted to pursue his passion outside the reach of Steve Plotnicki, he would have to put on an event himself.

With the Roskis out of the picture, Fierst inched toward forging a peace between the partners. Plotnicki recognized the progress and, at the start of 1999, agreed to take over paying Fierst's fee.

Fierst's first proposed settlement gave Thorpe the right to produce a live Robot Wars event in San Francisco, once a year. Thorpe would get $250,000 in cash for his 50 percent share of the Robot Wars business, plus health insurance and $40,000 in salary over the next seven months. The deal would also require Thorpe to rehabilitate the Robot Wars business in the eyes of the builders. Plotnicki signed on the dotted line.

Thorpe took a fresh look at the agreement and couldn't sign it. He didn't want to be in business with Profile any longer. Fierst recalled, "Marc did not trust Steve. Marc did not like him. Marc did not want to be in business with him. And Marc didn't want to be in a position where anything he said could be twisted and used against him in some nefarious way." Moreover, Thorpe was dubious that the robot builders would participate in any Profile-related event. They "considered Plotnicki to be the devil incarnate," Fierst said.

Thorpe rewrote Fierst's settlement proposal. He removed his commitment to rehabilitate the business. He eliminated his role as producer of the event and wrote on the new contract, "Both parties acknowledge that Thorpe has determined that for reasons associated with the dispute, few if any key competitors will participate in such an event within the present context." Then he signed it.

Plotnicki scoffed at this. Fierst had to come up with something else.

The second proposed agreement dropped Thorpe's participation in Robot Wars altogether. The creator of the sport would not produce any more events, or get a salary or health insurance. He would receive a single payment of $250,000, and a royalty of 10 percent of all future Robot Wars income. His percentage of the company would be transferred to Plotnicki, who would have complete control over Robot Wars. The partners would finally, irrevocably, go their separate ways. The only catch: Plotnicki still demanded that Thorpe agree to use "his reasonable, best efforts" to promote the Robot Wars property. The agreement said, "Neither party will make any disparaging comments about the other from this moment forward, and each agrees to use his reasonable best efforts to rehabilitate the reputations of the others to the extent that they have been damaged during the course of the parties' dispute."

This was essential for Plotnicki: If Robot Wars were to succeed, Thorpe would have to convince the community to compete. If he could drag them halfway across the state for a dull bankruptcy hearing, or organize a massive letter-writing campaign, Plotnicki reasoned, he could sure as hell get them and their robots to Fort Mason that August.

Another key part of the settlement required Thorpe to hand over all the entry forms from previous competitions to Profile. These were the contracts, signed by the builders, which gave Robot Wars the rights to make toys based on their robots. During the negotiations,

Thorpe disclosed to Fierst that he was missing the forms from the last Robot Wars event in 1997. Fierst said the settlement required him only to turn over what he had in his possession.

Trey Roski was opposed to the new deal and urged Thorpe not to sign it. "He's never treated you fairly in the past, and he's not going to treat you fairly now," he said.

But Thorpe felt he couldn't fight anymore. He was tired, he already owed Roski the $150,000, and legal bills were mounting every day. The stress on his family was extreme. "I was at the point where I had very few choices," Thorpe said. Even though he hated the deal and considered it a "surrender," on February 1, 1999, he signed on the dotted line and removed himself from direct involvement in the sport he had founded. A month later, the bankruptcy court approved the settlement. Plotnicki now had complete control of Robot Wars.

Dan Danknick, the builder of Agamemnon and Alexander, always wanted to build his own robot. Growing up in Orange County in the seventies, he pored over *Radio Electronics* magazine every month and immersed himself in a series of articles on constructing a home robot. Danknick's youthful effort ended in failure—he ran out of money to buy parts, but he never relinquished the dream. He attended college at the University of California–Irvine, got booted from the computer science program for taking a year off to work at an entertainment robotics company, and ended up graduating with a degree in applied physics. Later in the nineties, he worked as one of Disney's vaunted "Imagineers," a group that was "knocking together the next generation of animatronics control software for the new park in Tokyo," he said. To Danknick, Thorpe's competition rocked. He jumped at the opportunity to compete and fell irretrievably in love with robotic sports.

Like his fellow gearheads, Danknick rewired his life around his hobby. He pulled his car out of the garage, filled the open space with an oversize mill and lathe, and crammed cabinet shelves full of aluminum, steel, and plastics. Investigating every opportunity to pursue the pastime, he volunteered as a judge in the high-school robotics competition FIRST, and designed and built his own parts, which he sold to other builders on his popular website, TeamDelta.com.

In the year before Marc Thorpe's settlement in bankruptcy court, no builder heaped more scorn on the chief executive of Profile Records than Danknick. In frequent posts on Carlo Bertocchini's Web forum, Danknick called Plotnicki "weasel boy." He joked about the quality of the stationery used in the letters Profile sent to the builders. He called Mentorn CEO Tom Gutteridge "an official goon of the dubiously legal entity Plotnicki set up in the U.K.," and in 1998, when Gutteridge floated the possibility of staging a replacement event in the United States under a license from Profile, Danknick slammed the proposal and urged other builders to reject it as well. "Team Delta will not be part of any Mentorn production anywhere in the world," he wrote.

Danknick was among the builders who congregated in Santa Rosa court to watch Marc Thorpe's first bankruptcy hearing, driving up from L.A. and staying at Roski's house in Marin County. That night, Danknick and fellow robot builders Peter Abrahamson and Christian Carlberg sat in Roski's living room and played with his three enormous weimaraner dogs. Roski vented his frustration about his failed efforts to help Thorpe and wondered if the protracted litigation would ever end. He also floated a new possibility: Maybe he should just stage his own competition, under a new name, without Thorpe or Profile. His house guests were excited. "We all jumped at the idea," Abrahamson recalled.

For his part, Danknick knew that more than anyone else, the driver of La Machine yearned for competition to resume. "It was like his favorite amusement park had been burned down," Danknick said. "He was one of the few people in a position to fight."

Meanwhile, from his perch in New York City where he monitored the robot-building Web forums, Steve Plotnicki was beginning to notice Dan Danknick. "He was among the most obnoxious of the robot community, posting all types of crazy things on the Internet," Plotnicki said.

Danknick's perspective on the dispute suddenly changed right around the time Marc Thorpe signed his settlement with Plotnicki, after a phone call from fellow builder Tony Buchignani. Buchignani was the maker of the featherweight champion Wedge of Doom, and coincidentally, a lawyer at an L.A. firm doing work for Profile Records. One day, another lawyer at the firm, who happened to be friends with Plotnicki, walked into Buchignani's office and asked him, "Aren't you involved in all that robot-building stuff?" Buchignani said he was. The lawyer told him that certain builders were putting themselves in a position to be sued for slander—and mentioned Danknick. Buchignani immediately called Danknick to inform him that he stood on the edge of a legal precipice. Buchignani explained, "You have to understand how things work, Dan. You're damaging a property of someone who has a lot of money and legal resources."

Other builders had reacted to Profile's aggressive tactics in different ways. Carlo Bertocchini sent those unrepentant e-mails back to the lawyers at Delphi and got himself sued. Gary Cline canceled his event, Robotica, and slunk off. Trey Roski had the resources to fight back.

Danknick's response was altogether unique. His was a rational, analytical mind, and he reassessed the situation in the context of what was at stake—his financial security, his family, and his future.

The phone call from Buchignani, he said, was like "a shot to the head. I suddenly understood that freedom of speech only applies to governments, not to civil situations." Internet chatters who slam stocks in Yahoo forums get sued all the time; employees get canned from private companies for disparaging their supervisors or bosses. Resulting suits are often booted from the courts by the toe of the First Amendment, but for the little guy with few resources, simply mounting a legal defense constitutes punishment itself.

So Danknick reversed himself. He went back to delete his most vituperative, anti-Profile declarations on the Team Delta website and on the Delphi forum and carefully measured all his future posts. He also talked to Buchignani and decided to pursue the strategy that best assured he wouldn't be targeted: neutrality. Both builders decided to stay out of the fight, but do everything they otherwise could to forward the prospects of the sport they loved.

That spring, when Carlo Bertocchini finally agreed to give up control of the Robot Wars forum as a condition for having the lawsuit against him dropped, Danknick and Buchignani offered to take over as moderators. They worked out a deal with Plotnicki: They would monitor the site and delete any personal attacks against Robot Wars or Profile. In return, they could sell advertisements on the forum and keep the cash. The plan was, at least in part, inspired by the excitement of the dot-com boom: The proprietors of websites with frequent and loyal visitors were cashing in by selling Web banners. Danknick and Buchignani thought the Web forum could be a lucrative property.

The other builders puzzled over Danknick's sudden "neutrality." They couldn't understand how he had moved from vituperatively anti-Profile, to actually working with the New York music exec.

Carlo Bertocchini called Danknick and asked him this directly. "I don't have the financial resources to fight if I get sued," Danknick explained. Bertocchini told him not to worry and that "Trey will help defend you." Danknick said he didn't want to rely on the theoretical generosity of someone else. Bertocchini couldn't understand how working with their common enemy constituted a neutral solution to the problem.

"Carlo really never spoke to me after that," Danknick said.

Word of Marc Thorpe's new settlement was leaked onto the builder's forum on the net. One post described Plotnicki as someone who had "squeezed every last bit of ownership out of Thorpe" and was now "suing him again to make him dance like a puppet." Another post, written by Gary Cline, was titled, "Which of you would be a prostitute?" referring to anyone who would compete in a Profile-related event.

The messages were deleted by the forum's new moderators. Danknick and Buchignani were acting on requests from Profile Records in New York. The robot builders grumbled about censorship.

Meanwhile, there were rumors about upcoming events, but no hard information. Eventually there were stiffly worded press releases.

The first one came from Profile Records, announcing a new event, Robot Wars '99, to be held that August, over the weekend of the twenty-first and twenty-second. "I know that the U.S. robot makers have been anxious for the event to be held. Now they will once again have their chance to compete during the summer of 1999," Marc Thorpe was quoted as saying in the press release. The following paragraph quoted Steve Plotnicki: "Now that the litigation is behind us, we can finally get the property back on track in the U.S." Thorpe, it said, was "to be connected to [Robot Wars] as its roving ambassador."

The press release also dangled an enticing possibility to the builders: Robotic combat in the United States might be *televised.* "Discussions are under way to broadcast this summer's event," Plotnicki was quoted as saying.

For Steve Plotnicki, all that was left to complete his triumph was for Thorpe to fulfill his side of the agreement and drive the point home in the community of robot builders. Gary Pini called Thorpe and asked him to post a message on the Delphi Web forum, emphasizing his support for the Robot Wars event.

Marc Thorpe was now facing a dilemma. He had entered into a legally binding agreement to promote the Robot Wars property. Emotionally, though, he couldn't. Rehabilitate Robot Wars! If anyone needed rehabilitating, it was him, and the quivering right side of his body. Now he was basically being forced to lie. He was required to speak with someone else's voice, for someone else's agenda. He had gone from straining desperately to preserve control of his creation, to losing control not only of the sport, but of himself, his own voice, thoughts, and opinions. It ran horribly afoul of the gearhead sensibility.

That week, he logged on to the Robot Wars Web forum and posted this:

> I informed Gary Pini of Profile Holdings, which now owns Robot Wars, that when people ask me what the situation is, I tell them that I am no longer an owner but stand to benefit from the success of the business as a royalty partici-

> pant. Regarding participation in Robot Wars '99, I tell people that I want them to do what they genuinely want to do and that I will respect their decisions regardless of what they are. I do have an obligation as part of the settlement to promote the business and I intend to do so when interviewed by the press or if I appear in public. Having created the sport and having seen the wonderful community that has grown around it, I will always have good things to say about it. For clarification, insofar as Robot Wars '99 is concerned, Profile is financing and producing the event and I will not be participating in the productions. Circumstances permitting, I will be there. In the meantime, I wish Profile the best of luck and success with it.

The following day, Trey Roski sent out a press release announcing another event, "Battlebots."

This was the idea Roski had suggested to his house guests on the night of Marc Thorpe's failed bankruptcy auction, and the idea he unveiled to other builders on the day of the underground street fight. "We were never going to get anything from Plotnicki," Roski said. "The sport needed to be done right," with royalties from a TV contract and toys flowing back to the hobbyists who were putting their time and money on the line.

Roski convinced his cousin Greg Munson to close Impact Media and join him in creating a new robotic combat company. Munson would plan and organize the events, while Roski would head the business and direct the effort to get Battlebots on television and toy store shelves. He would have to work around the limitations of his dyslexia, getting others to dictate his letters and read contracts to him.

"For those of you who don't know me," said a letter Roski posted on the new Battlebots Web page, "I'm the guy who drives La Machine. I have competed in robot combat events since 1995. The sport of battling robots totally consumes me—I love it! Since my first competition, I knew this is where I had to be. Like most of you, it's in my blood."

Roski went on to explain that his family owned parts of the L.A. Kings hockey team, the L.A. Lakers, and the Staples Center arena. "All of us have waited long enough," he said. "It's now time to get back to our passion: Battling Bots." He set the first Battlebots competition for August 14 and 15 in Long Beach, California, one week before the dates previously set for Robot Wars '99.

Battlebots! This was what the competitors had been waiting for, an entirely new event free from the stranglehold of the (to their mind) villainous Plotnicki. Roski pledged that the builders would receive 15 percent of all merchandising revenues, plus a cut on any TV broadcast that featured their robots. Todd Mendenhall quickly initiated a new Battlebots Web forum on the Delphi service, and most of the builders migrated over. The Robot Wars forum would eventually be completely deserted, dashing Danknick and Buchignani's hopes to sell ads.

Everyone was thrilled, except for Steve Plotnicki and Gary Pini. They went into crisis mode. Pini cut short a vacation in Miami and flew back to New York, while Plotnicki called his attorneys and began planning his counterattack. It was happening all over again—just like Run-D.M.C. and Cory Robbins. His direct enemy lost the control battle, and an unholy ally emerged from the shadows to take up the fight.

"It's how it comes out in the end that really counts," Plotnicki had said, and this was far from over. Most of these builders had strong emotional connections to Marc Thorpe, and in his opinion, all Thorpe needed to do was fulfill his responsibility under the settlement agreement by convincing the builders to come to Robot Wars instead of Battlebots.

Thorpe got a phone call from his lawyer after the Battlebots press release went out. His attorney, William Weintraub, told him that Plotnicki believed he was secretly colluding with Roski in violation of the settlement. Thorpe maintained he didn't know a thing about Battlebots until it was announced. Weintraub told Thorpe that he should do his best to direct the builders to the Robot Wars event, or he'd be in violation of the settlement agreement.

Now Thorpe truly was alone. Roski, his friend, was staging another event, and Thorpe was obligated to back Robot Wars, over which he had no control. On the line was $250,000 in hard cash—what Profile owed him under the agreement—and thousands more in royalties. It represented the financial salvation of his family and security for the uncertain future.

All he needed to do to earn that bounty was go online and direct the builders to Robot Wars. He could have lied and told his followers he wanted them at Robot Wars; that he thought Profile was going to be a good steward. Conceivably, he could have also called Roski and begged him not to put on the rival event.

But taking that path was impossible for Thorpe. He couldn't lie to his friends within the builder community. He was an artist and a gearhead, and this combination of skill and obsession connotes an unshakable integrity and stubbornness. He took to his computer again, and what he wrote on the Delphi forum would plague him in the courtroom for years to come.

> I want everyone to know that I have nothing to do with Battlebots or Trey's business in any way. He is a dear friend and I wish him well with all he does, but I have no involvement at all in Battlebots. The first I heard about it was the day of the press release. Second, I have also said that I wish Steve Plotnicki the best of luck and success with the event that he is planning this summer. I say this not simply because I am looking forward and not back, but because as I said in a previous post, I stand to benefit from his success. However, according to Steve's attorney and Gary Pini, and in a letter that I received today from Steve, I am not living up to my obligations within the settlement agreement. Specifically, they say that I am not doing enough to get robot builders to participate in Steve's event and to rehabilitate his reputation. So . . . once and for all, imagine me opening up my front door, standing with hands cupped on either side of my mouth, and shouting at the top of my lungs for all to hear, "STEVE PLOTNICKI IS A NICE GUY AND I WANT EVERYONE FROM BILL CLINTON TO C3PO TO COME TO ROBOT WARS '99 BECAUSE IT WILL BE

> THE BEST EVENT EVER!" Hopefully this will satisfy all concerned and I can go back to rebuilding my life . . . unless the settlement agreement requires me to come back with a "DIE BATTLEBOTS" post.

He signed off by comparing himself to LAPD beating victim Rodney King: "Can't we all just get along?—Rodney Thorpe."

Plotnicki was furious. He demanded that Thorpe contact each individual robot maker and urge him to attend Robot Wars. Thorpe resisted, insisting he didn't have enough time to make 70 phone calls, and that he had never hyped Robot Wars so flagrantly in the past. Plotnicki wrote a statement in Thorpe's name, which expressed enthusiasm for the event, and demanded that he sign it. Thorpe rewrote the statement, but couldn't bring himself to send it out. Then Thorpe himself wrote an enthusiastic press release about Robot Wars and sent it to Plotnicki with a note saying it was a glimpse of things to come—if Plotnicki dropped all the legal claims against him. Plotnicki refused and told Thorpe to send it out anyway. And so on and so forth.

That July, Thorpe at least partly caved in to the pressure. He posted a message to both the Robot Wars and Battlebots Web forums, claiming he was now actively involved with the production of the Robot Wars event, and urged builders to attend it over Battlebots. "If you find that you cannot attend more than one event, I would like to encourage you to attend Robot Wars—the event that has special significance to me because I conceived it and it follows the rules that I developed. I will be at Robot Wars '99 and I hope to see you there."

But Thorpe still couldn't win. The builders recognized his change in tone and skeptically wondered about it. Jim Smentkowski called Thorpe at home, then, after an awkward conversation, posted to his website, "I barely got any information from Marc, and he sounded extremely exhausted the whole time. When I asked about the posting, the only response I could get from him was 'it speaks for itself' and 'I'd rather not talk about it.' Needless to say, this seems a bit strange to me. Knowing I wouldn't really be able to find out what was going on, I asked Marc how he was doing, to which he replied, 'I'm doing the best I can.'"

In April 1999, Steve Plotnicki filed the predictable lawsuit against Battlebots in New York federal court, asking for a preliminary injunction to stop the event, just as he had with Robotica the year before. The suit named Trey Roski and, to the younger Roski's chagrin, his father, Ed, who had been dubious of his son's involvement in Thorpe's legal problems from the beginning. Plotnicki alleged that the defendants, "With the aid of their sophisticated legal counsel and others, devised a scheme and provided financing to: a) prevent Robot Wars from conducting business; b) usurp Robot Wars goodwill and destroy its reputation; and c) commandeer the valuable Robot Wars mark."

In the 50-page lawsuit, Plotnicki said that Thorpe's "bad faith bankruptcy" was designed solely to break his agreement with Profile. He argued that Battlebots was putting on the same event as Robot Wars, with the same robots, the same builders, and tellingly, the same format and rules. It wasn't a case of someone starting an unrelated robot competition with another name; Plotnicki felt Roski was putting on *his* robot competition under another name.

In Robot Wars, the head-to-head battles were called "face-offs," while according to the rules posted on the Battlebots website, the new competition called them "duels."

In Robot Wars, the free-for-alls at the end of the competition were called "melees." Battlebots called them "rumbles."

Robot Wars matches lasted five minutes. So did Battlebots'. A robot that fragmented into smaller bots in Robot Wars was called a "clusterbot." In Battlebots, it was a "multibot."

Plotnicki thought Battlebots was an obvious rip-off—just like Gary Cline's Robotica the year before.

Fearfully, Trey Roski told his millionaire father that he had gotten him sued. To his surprise, the elder Roski didn't care. "He just seemed to think it was a cost of doing business," Roski says. "I was more nervous about it than he was."

The Roskis' lawyers met Plotnicki's 50-page complaint with voluminous filings of their own. In long, telephone-book-size affidavits and exhibits, they pointed out that their rule book, written by SORC head Todd Mendenhall, represented a greatly expanded and altered version of Marc Thorpe's original guidelines. They denied that Thorpe had anything to do with the formation of Battlebots, and they defended Battlebots by asserting that robotic combat was a sport, not a "property." Just as football did not belong solely to the NFL, robotic competition was not the sole intellectual property of Robot Wars LLC. To support this claim, they trotted out descriptions of the annual FIRST competition for high-schoolers, Critter Crunch from the Mile-High science-fiction convention in Denver, and robotic sumo in Japan.

The parties met on June 7, 1999, in the courtroom of Judge Jed Rakoff, in the same building as Judge Batts's courtroom in downtown Manhattan. There were no robot builders present this time, only the two high-priced legal teams for Profile Records and Edward Roski's Majestic Realty. The hearing was brief, and the judge took the matter under submission. He would decide whether Battlebots constituted a flagrant violation of trademark, or if it had a right to proceed with its August event.

Nine days later, Rakoff ruled: Profile's request for a preliminary injunction to stop Battlebots was denied. If Battlebots was harming Robot Wars, Rakoff decided, damages could be tabulated at a trial, to be set later.

The competitors were jubilant. Once again, they thought they had won, that the mighty Profile Records dragon had been slain. The court's decision "really made me Sm:)e," joked builder Derek Young in one post.

Battlebots had a green light to proceed. Fifty-six builders signed up to compete in the inaugural Battlebots competition in Long Beach. Robot Wars, to be held one week later in San Francisco, got only 21 entrants. It wasn't enough to stage a competition, and Steve Plotnicki canceled the event. Robot Wars in the United States appeared to be finished.

Plotnicki believed he had been swindled: that Marc Thorpe, conspiring behind the scenes with Trey Roski, had deliberately killed Robot Wars. He dropped the lawsuit against Roski and his father, but it was only to come back with a stronger assault in front of a more amenable judge. "It's how it turns out in the end that really counts," he said.

Team La Machine was directing the show now, and a call for help went out to robot builders. Like Robotica the year before, the new event would be a community effort. Builders like Carlo Bertocchini, Jim Smentkowski, and Scott LaValley gathered to build and paint the components of the new arena and ship them to Long Beach. Pete Lambertson, the San Francisco machine shop owner who had helped fabricate La Machine's curved plow in 1996, was hired to design what would be called the "Battlebox." It was 48 feet long and wide and 22 feet high, made of one-inch-thick Lexan polycarbonate, and it cost about $100,000 to build. The previous Robot Wars arena had been only eight feet tall and made of one-quarter-inch plastic. Trey Roski wanted to make sure the crocodiles were safely contained. He also wanted to give the builders more of a chance to compete, and made the sport double elimination.

Roski rented out the landmark Long Beach Pyramid south of Los Angeles, on the campus of his old alma mater. It seated 5,000 spectators, but Battlebots sold only a fraction of those seats. Unlike the San Francisco Bay Area, the population of Southern California was not yet familiar with this new pastime of fighting robots.

The builders were thrilled to compete anyway after two long years. Carlo Bertocchini brought Biohazard, Gage Cauchois was there with Vlad the Impaler. Mike Winter and his daughter, Lisa, drove down from Wisconsin with Tentoumushi, a lightweight robot equipped with (of all things) a red, ladybug-shaped sandbox. The inverted sandbox, attached to a pair of wheels, could be lifted and dropped onto opponents, smothering them against an electric saw that was concealed underneath.

Jonathan Ridder brought his new spinner, Ziggo, and Stephen

Felk brought his heavyweight, Voltarc. Finally, they would get a chance to compete inside the ring. Glued to the base of Voltarc was the 1941 nickel Felk had found outside Fort Mason three years earlier. It symbolized the presence of his father.

Also present were two British artists, Ian Lewis and Simon Scott, and their creation, Razer. The 32-year-old high-school chums from Poole, England, had participated in the recently taped second season of the U.K. Robot Wars game show and won the award for best design. Razer was an elegantly designed mechanical scorpion: Its hydraulic-powered beak lifted up, then punctured opponents with tremendous force, opening them up like a can opener.

Jim Smentkowski had one of the most impressive new robots, Nightmare, essentially a large toothed vertically spinning disc. Two heavy blocks of steel were welded at opposite sides of a 40-inch diameter aluminum blade, and a four-horsepower motor spun the blade upward, reaching speeds of 300 miles per hour at the blade's tips.

Marc Thorpe also drove down for the event. He got lost on the way and had to stop at a succession of convenience stores to get directions. This time, he was a spectator, and he was using the opportunity to try to satisfy one of the provisions of his settlement. That weekend, Thorpe wandered the pits at Battlebots, asking the veteran builders to re-sign the entry forms from the 1997 competition that he had lost. Less than half of them did. Trey Roski took his sign-up form and wrote "NO" on it in giant letters. He would later have one of his father's lawyers send a letter to Profile, threatening litigation if any Robot Wars literature used La Machine's likeness.

That weekend, against a booming backdrop of techno music and hip-hop, Ziggo went up against Jason Bartis's Dr. Inferno Junior and sent a hail of shrapnel into the walls of the Battlebox. Then Nightmare performed a similar feat on Patrick Campbell's new hammer-bot, Frenzy. Roski had hired much of the same safety and production crew that worked on Robot Wars, and once again, they were worried about the most destructive robots. Despite all the work done to build the powerful Battlebox, new robots like Ziggo and Nightmare might hurl shrapnel over the 22-foot-high walls. So they asked Ridder and Smentkowski to moderate their weapons. Ridder lowered the voltage on his motor to make Ziggo spin slower, and Smentkowski changed the direction of his weapon blade. Instead of spinning up toward the ceiling, it spun down to the floor. This was safer but less effective, and Nightmare lost its next match to the wedge-shaped Punjar.

Bertocchini's Biohazard performed its customary romp through the heavyweight opposition. It won all five of its matches and took first place in the heavyweight class. Its titanium armor was barely scratched.

Stephen Felk's Voltarc lost its first two matches. Felk was depressed, having worked for three years, only to promptly lose. But in the heavyweight rumble he positioned his bot's articulated lifting arm underneath an interlocked Razer and Frenzy, and hoisted both robots into the air at the same time. As the crowd cheered, Voltarc dragged them across the arena.

Even better, a CNN crew was on hand and interviewed Felk for a story about Battlebots. From his L.A. hotel room the next day, he watched himself on global television, evangelizing for the new sport he loved. "I have a sculpting background, and I sort of have that engineer's mind," he said on the broadcast. "I love the competition of it, I love being involved in an event like this. And when I first saw it, I was gone." If he had any thoughts about giving the whole thing up, they were vanquished by the exhilaration of seeing himself on television.

Concluding the weekend was a surprise from Marc Thorpe's former ILM colleague Mark Setrakian. The maker of The Master and Snake showed up with a new robot, Mechadon. The special-effects master had outdone himself: It was a 475-pound walking crab that tiptoed about on the sharp alloy points of six massive pincer-legs. There were three legs on each side, attached to three independently moving segments of the main frame. The beast had 14 separate motors. When Mechadon turned and writhed on the ground, it looked like some kind of insectoid Rubix cube. Setrakian, who built the Mechadon in four largely sleepless weeks, demonstrated it to an awed crowd, but it was too heavy to actually fight.

For Thorpe, the entire weekend was a jumble of mixed feelings. He loved seeing the community back together after the long hiatus, but begging for signatures and watching from the stands was strange and uncomfortable. Then there was the horrible announcer. To MC the event, Roski had hired a former participant of the syndicated TV show *American Gladiators* named Lee Reherman. Reherman's muscles bulged from his black tuxedo; his neck was thicker than a few of the robot builders' waists. He didn't just announce the matches, he belted them out in a high-testosterone, over-the-top, energetic roar, slowing down only long enough to savor the possible phallic references in his descriptions of the robots.

"Behind me right here is the BattleBox!" he screamed to the CNN cameras that weekend. "Once you get into the Battlebox, the rules are really complicated. The rules are . . . there are no rules!"

The builders were embarrassed, but Thorpe was appalled. He had prided himself on avoiding the aura of professional wrestling with Robot Wars. He had cultivated a more serious approach, inviting robot artists like Mark Pauline and Clayton Bailey to present their art. While Reherman barked to the crowd, Thorpe snuck up behind Greg Munson and bristled, "You've got to get rid of that guy!" But Munson didn't do anything. Thorpe wasn't in charge anymore.

Like Marc Thorpe before him, Trey Roski knew that making robotic combat work as a business was a tricky proposition. The event relied solely on the goodwill of the gearheads, mechanical obsessives willing to spend the time, money, and sweat equity to build the robots necessary for competition. Anger the builders and the event could crumble, as Steve Plotnicki had found out.

However, Roski figured that if he delivered a solid, safe venue for competition, and promised financial reward, the builders would stay faithful to his event. He pledged that Battlebots competitors would get 15 percent of any toy revenue—double what the contracts for Robot Wars had delineated, even more than the 10 percent that Mentorn was promising the British robot builders who participated in the early seasons of the U.K. TV show.

Roski invited representatives from the Hasbro and Mattel toy companies to the Long Beach Pyramid, but he knew that to really convince the toy makers that he could move merchandise, television had to implant the concept of fighting robots in the public mind.

For help with this, he turned to Lenny Stucker.

Stucker was the cofounder of TalentWorks, a 15-year-old pay-per-view TV production company based in New York. Stucker looked a bit like Steven Seagal: tall, slicked-back hair, impeccably dressed, with the solemn air of someone you don't want to cross. Stucker was a veteran of the TV business and had produced World Series and Super Bowls for NBC for 11 years. His partner, Rob Beiner, had worked with Howard Cosell at ABC. They claimed to have 22 Emmys between them.

In the nineties, Stucker had seen Marc Thorpe's Robot Wars on a TV news program and loved it. He called Steve Plotnicki at Profile Records and suggested they meet to discuss a relationship. TalentWorks, which specialized in pay-per-view, wanted to acquire the international broadcast rights for five years. Stucker offered to pay all production and development costs for 60 percent of the profits.

They couldn't strike a deal. Plotnicki thought it was strange that Stucker offered to send a limo to bring him to a meeting; Stucker thought it was odd that Plotnicki wanted to see TalentWorks' financial information before he would consider working with the firm. The parties emerged from their unsuccessful dialogue disdainful of each other.

Still, Stucker thought robotic combat had potential. "For a sport, it has all the puzzle parts. Everyone can play. No one gets hurt. And it teaches engineering," he said. Moreover, the British TV version of Robot Wars was getting good ratings on the BBC. Stucker knew robotic combat would play well on television in its birthplace, the United States.

When Stucker found the Battlebots website in 1999, he recognized a second chance to get involved. He went to Long Beach to watch the first competition, seeing Razer, Mechadon, Voltarc, and the others. Afterward, he contacted Trey Roski and invited him to his office in New York. Roski's father happened to be in town the same day. Father and son visited TalentWorks and liked what they saw.

Stucker, Roski, and Munson began shaping the sport for television: They shortened the match length from a potentially dull five minutes to a snappier three and moved from audience judging to an actual panel of judges, formulating a point system that tallied aggression, damage, and strategy. They installed in the Battlebox a "Christmas tree," a traffic signal–like mechanism borrowed from drag racing, to indicate the start of a match in a visually exciting way. Stucker brought in a professional ring announcer, Mark Beiro, known primarily in the boxing and wrestling worlds, to emcee events in place of the bombastic American Gladiator. Finally, they talked about getting rid of the double-elimination format of the first Battlebots event. "No sport gives its players two chances," Stucker argued.

Roski and Munson resisted that last change. They were competitors themselves and knew the pain and disappointment of spending thousands of dollars on a robot, preparing for months or years, only to get knocked out in an unfortunate instant. Stucker convinced them it was necessary for the sport to work on television. Eventually, they compromised.

On two points though, Roski refused to budge: He insisted that the builders get a 15 percent cut of the television revenues, and he demanded that anyone who showed up with a safe, workable robot that conformed to the guidelines could enter the competition. It wasn't just the largesse of the rich or a strategy to capture the builders' allegiance; he wanted the sport to be able to produce for others what it had done for him. "Look what I got out of it," he said. "I was nothing, out of shape, overweight. I had a chance to go up on equal grounds against all these brilliant guys. It's special. It belongs to contestants who want to play the game."

Four months after he signed Battlebots as a client, Stucker got the sport a spot on the InDemand Pay-Per-View service. A new competition was scheduled for November 1999, at the All American Sportspark in Las Vegas. The competitors would stay at Silverton Hotel and Casino, owned by the Roskis' Majestic Realty. The event would feature only heavyweights and a new class, super heavyweights, created specifically so that audiences could see Mark Setrakian's creepy, crawly 488-pound Mechadon in action. Dan Danknick heard Battlebots was looking for a technical correspondent for the on-air broadcast. He volunteered, and got the gig.

The builders went to work preparing for the new event. Finally, they were going to be on television, the exalted stage. Christian Carlberg in Los Angeles built a new robot, Minion. Trey Roski redid his own robot, Ginsu, adding two more carbide-tipped blades as wheels. Stephen Felk revamped Voltarc, welding shut the rotating wheel shields on the top of the bot that weren't working very well and adding a hinged scoop to the back of the wedge that could self-right his contraption if it got flipped. All the bots were getting better, stronger. The engineering ethos was at work: Adapt or die.

There was, however, an exception: Jamie Hyneman brought Blendo, but it was basically the same whirling-dervish metal wok that he had entered in Robot Wars in 1995. The pits this time were located outside, in a cold, drafty tent next to the SportsPark, and a screen inside the tent simulcasted the action from the arena. The pits got very quiet when Hyneman brought his feared, mythic spinner into the Battlebox, because the builders knew Blendo wouldn't get disqualified this time. The new Battlebots arena could withstand the damage. There was even a plastic roof that prevented debris from clearing the 22-foot-high walls.

Blendo's first bout was a rematch against Punjar. Hyneman still believed he had the most powerful contestant around—he still wanted to "spray another robot around the arena," he said. But since his devastating, one-shot loss to Blendo in Robot Wars '97, carpenter Ramiro Mallari had upgraded his wedge-shaped bot with heavier armor and a more compact motor-battery drive box. It was designed to take a beating: Any weapon that punctured the aluminum shell would have to travel almost a foot before it reached a sensitive component.

As Hyneman lined up Blendo across the Battlebox from Punjar, all the builders in the pits gathered to watch the match on the big TV screens. Seventeen hundred spectators shifted to the edge of their seats. As in Long Beach, the population of Las Vegas didn't know robotic combat; the organizers had to comb streets and local colleges to fill the arena. But now the crowd was into it.

The match lasted only a minute. Mallari raced his refurbished wedge to the other side of the box and plowed into the whirling Blendo. The robots collided, then spun away from each other, whiplashed by Blendo's rotational energy. Blendo was still moving—but this time, so was Punjar. Mallari straightened out his wedge and bounced it off Blendo once more, pushing the spinner against the wall. The great metallic wok started to slow down. Again Punjar crashed into it, and again the robots were flung away from each other. Blendo was driven up the slight incline that led to the Battlebox doorway. Slowly, incredibly, the great destructive spinner stopped spinning and then stopped moving altogether.

With a pained expression, Hyneman gave the referee the signal for the tap-out. Ramiro Mallari started to jump up and down as if he had won the lottery. In the pit tent outside, the competitors went wild: After four years, Blendo had finally been defeated.

¤

There were other upsets that weekend: Biohazard lost to the pneumatic forklift of Vlad the Impaler, its first-ever defeat in one-on-one competition. Cauchois was able to spear one of Vlad's sharp spikes underneath Biohazard's titanium skirt, then pop the forklift up and drag Biohazard helplessly around the arena. Stephen Felk enjoyed his first string of victories as well, avenging his loss to the British builders of Razer, then beating their compatriots from Cambridge, makers of the robot Mortis. Voltarc made it to the finals before losing to Vlad the Impaler. The next day, Bertocchini and Biohazard returned to win the award in the heavyweight rumble.

Super heavyweights Mechadon and Ginsu fought an otherworldly match—they looked like two aliens wrestling against the polycarbonate wall. Mechadon's heavy pincers surrounded Ginsu like the tentacles of an octopus, bending Trey Roski's bladed wheels. Even though Ginsu cut a motor out of Mechadon, Setrakian's bot still had 13 others. It continued to writhe until Ginsu flopped over onto its side and couldn't move. Mechadon was too damaged to fight again, so Christian Carlberg's six-wheeled, blade-mounted Minion won the super heavyweight category. The 29-year-old L.A.-based special-effects worker was now thoroughly addicted to the sport.

At the conclusion of the event, Roski presented each winner with one of Battlebots' new "golden nuts," 20-pound silver trophies in the shape of a real hardware nut. Then it was time for Roski and Greg Munson to take their turn in front of the cameras, as the spokespersons for the new sport. Like Marc Thorpe before them, they had to answer The Question, the one that everybody asked—was robotic combat violent?

"This is violent and it's fun and it's damage, but it's only robots, humans never get hurt," Munson answered.

"It's very dangerous and destructive," Roski said. "But on the other hand, it's safe for the family. It's safe for the audience. It's safe for everyone to come and enjoy all this destruction."

When the cameras were turned off and most of the spectators had filed out of the arena, the builders presented Roski and Munson with their own award. It was a plaque, reading, "In our sport's darkest hour, you have shown a guiding light."

Chapter 7

The Exalted Stage

In the future, I believe that even under ideal circumstances that Robot Wars' relationship with the robot community will at times be not unlike the strained player-owner relationships in other sports. . . . The more successful the sport, the more money there is to argue about.

—Marc Thorpe, from a 1999 e-mail

The three Robobabes appeared on rollerblades.

They wore tight black pants, midriff-baring black tops and yellow kneepads. They had streaks of black and yellow war paint on their cheeks, and they skated around the London Arena rapping about robots over headset microphones. Each was a professional dancer and that distinctly English phenomenon—sexpots who aren't very pretty.

It was the summer of 2001. More than 5,000 eight- to 12-year-old boys and girls tugged on the shirtsleeves of their mums and dads, crammed into the coliseum in London's Docklands district, eager to glimpse a live version of the show they watched every Friday night at seven: *Robot Wars.* This was a traveling Robot Wars tour, a sort of robotic ice capades, a hokey staged rendition of one of the most popular shows on BBC2. The kids, however, endured the Robobabes with a quizzical silence.

But then the Robobabes rolled away, and popular *Robot Wars* TV host Craig Charles announced the arrival of the house robots. This was what the kids were here to see—the stars of the weekly program. The house robots, built by the BBC visual-effects department, patrolled the *Robot Wars* arena each Friday night, entering the action to dispose of one or both of the competitors at the end of a fight. Their presence ensured a spectacular climax to every battle. They wielded drills, flamethrowers, chainsaws, and diamond-edged axes and weighed more than any competitor, up to 616 pounds. In the London arena, the kids greeted their introduction, one by one, with a full-throated roar.

"He might be a lightweight in house robot terms, weighing in at 105 kilograms, but he's a heavyweight when it comes to havoc. It's . . . SHUNT!

"She may not be able to waltz like her Australian namesake, but she does have hydraulic tusks. Truly a sting in her tail! It's . . . MATILDA!

"He's the flame-filled, militaristic machine of war . . . It's SERGEANT BASH!

"His weapons consist of pneumatically driven pincers, designed to immobilize all opponents, and a circular saw turning at over 3,000 rotations per minute. It's . . . DEAD METAL!

"Finally . . . the bot you've all been waiting for . . ." (The kids knew what was coming and infused the arena with a rattling roar.) "He's the scariest robot on Earth, weighing in at a massive 280 kilograms and standing 1.3 meters high. He's armed with a rotating drill and a fearsome hydraulic cutting saw. It's . . . SIR KILLALOT!"

The kids loved it. They wore *Robot Wars* T-shirts, caps, and sweatshirts and waved their *Robot Wars* foam-fingers frantically in the air. Seven million people watched the program in 17 countries, and here in the London Arena was the heart of that fan base, thousands of house robot–loving boys. The British TV show, which debuted in 1998, was a genuine sensation. Now British production agency Mentorn was taking a live version of the show on tour to eight different coliseums in the United Kingdom. Tickets ranged from 10 to 17 pounds, with discounts for families.

Midway through the performance, a new robot, Lightning, was introduced. It was a purple wedge, with silver electric bolts painted on either side. The front scoop was a pneumatic flipper, meant to pop other bots onto their backs. But this was no house robot; this was produced by a team of genuine garage gearheads from Essex, England, and they were here to audition for the fifth series of the TV show, to be taped at a London studio later that summer.

Lightning's makers were dressed up as mad scientists, complete with white lab coats, fake teeth, and wigs. It was meant to be a unique, goofy act, an attempt to stand out. They were among the 3,000 applicants who had applied for the show, and they were in the minority who got Mentorn's approval to build a robot. That didn't guarantee them anything, though. Only some 100 machines would be invited to the taping, many of them returning veterans. Lightning, the result of six months of work over evenings and weekends, would have to prove itself here, in front of a live audience.

For its trial match, staged after the introduction of the house robots, Lightning faced one of the fiercest bots from previous *Robot Wars* events—Razer. Ian Lewis and Simon Scott, whose scorpionlike robot so impressed Marc Thorpe at the first Battlebots event in Long Beach, were accompanying the house robots on this leg of their summer tour.

The makers of Lightning didn't stand a chance against the veterans, of course. Lewis drove Razer, while Simon Scott controlled its deadly hydraulic pincer. Razer raced across the arena, dodged Lightning's flipper, and put five puncture holes into the wedge, quickly disabling it.

"Pit! Pit! Pit!" the kids yelled out in unison. They were calling for the destruction of the loser, just as they were used to seeing on television. After a few moments, the house robot Sergeant Bash emerged from its corner of the arena. It took Lightning up in its front-mounted claw and dragged it to a cavity at the center of the floor. Lightning disappeared from view and a cloud of vapor rose in its place, to the complete delight of the young crowd.

Later, the Lightning team would get the bad news that their robot hadn't qualified for the filming of the TV series. Team leader Laurie Calvert tried to look at the bright side. "In one way, I'm pleased because this gives us time to develop and get competitive rather than be cannon fodder for another robot."

After the crowd of sated youngsters and weary parents streamed out of London Arena, the performers cleaned up in the cavernous area behind the stage. That afternoon they would do another show, then they would move across town to Wembley Arena for two more, with new auditioners serving as bait for the veteran robots and the raucous, paying crowds.

Ironically, Tom Gutteridge, Mentorn's CEO and the first man to turn Robot Wars into a television franchise, never thought that live robotic combat—like the kind Marc Thorpe pioneered—could work. "It's almost impossible to have a live event make money," he once said. Now, he was proving himself wrong. Mentorn was crisscrossing the country with the house robots, exploiting the popularity of the TV show and taking in 1.5 million pounds in four weeks. The events, Gutteridge bragged, were "about as commercially successful as you can get."

Making the *Robot Wars* TV pilot back in 1995 hadn't been easy. Tom Gutteridge's deputy, Steve Carsey, spent more than four months trying to recruit teams to build robots. "Every university in the country was contacted," he recalled. "Pretty much every industrial giant in the U.K. from Rolls Royce to British Aerospace was approached and certainly every radio-control model organization." As the filming date neared, organizations discovered the technical challenges and resources involved and pulled out. Only with the help of the U.S. builders and their robots—La Machine, Thor, and The Master—was Mentorn able to stage a show and get the executives of the BBC to acknowledge the concept's promise.

But that was only the beginning of Mentorn's struggle. While the TV execs smiled and enjoyed the action, the BBC itself was hesitant. "It was a very risky show," Tom Gutteridge said. "Nobody could see how it was going to work on TV." Mentorn spent two years designing sets, commissioning the house robots, and finding enough English *anoraks* to make for an adequate program. "There were scarcely 60 builders worldwide who could do this with any real credibility," Gutteridge said. "There were certainly not enough here in England to make a TV series."

In late 1997, the BBC finally ordered six episodes. Mentorn had found only 33 English teams to compete. To round out the group, Mentorn brought in three ringers—Carsey preferred to call them "reserves"—built by BBC special-effects hand Derek Foxwell, who officially became the *Robot Wars* safety coordinator.

In that first series, filmed in November '97 before a live audience, the 36 robots went through a kind of robot decathlon. First, players negotiated "the gauntlet," directing their bots through a maze while avoiding the patrolling house robots. Next, the builders competed in games like "sumo basho," where they tried to keep their bot on a raised platform as a house robot attempted to nudge them off, and "soccer," with the bots trying to push a spiked ball past a house robot goalie. Only in the final stage did the competitors face off against each other. The house robots (minus Sir Killalot, who came later, in Series Three) waited nearby to toss losers into the pit or hold them over the bath of flames that emerged at opportune times through a grate in the floor.

After the taping, Gutteridge got what he thought was terrible news. *Robot Wars* was going to air on BBC2 in what he called "the death slot" on Friday night, up against the popular talk show *TFI Friday.* He assumed *Robot Wars* would suffer a quick extinction, but by the second episode, ratings for his show had matched those of its rival, and soon surpassed them.

The TV series was a hit. Four million Brits watched that first series every Friday night at six. "It's a sanitized form of violence," Gutteridge told *The London Times* that year. "It's violence where no one gets hurt. It's like being in a cartoon."

Gutteridge traced all the success that robotic combat enjoyed afterward back to that first series. "I know it all started in San Francisco with Marc Thorpe. But actually it started in February 1998, when this phenomenon was exposed to four million people, and the word spread that people loved this."

The BBC loved it too, ordering up another 15 episodes. Now Mentorn found it easier to recruit roboteers for the show. After the first season, 3,500 people joined a *Robot Wars* mailing list on the Internet. Mentorn got thousands of calls from England's mechanically inclined, asking for more information about getting on the program. These were people like George Francis, an electronics engineer from Ipswich whose black, Lexan-sheathed Chaos 2 would use its powerful pneumatic flipper to send other bots up and over the side of the arena, and twins Dave and Derek Rose from Cambridge, who with their dad, Ken, would build Hypnodisc, an aluminum box that carried a massively spinning flywheel like an hors d'oeuvres plate. Just like Blendo in the U.S., Hypnodisc would fire pieces of shrapnel off the other robots into the arena walls.

It was great television, because it was made specifically for television, as opposed to the live tournament format of Marc Thorpe's Robot Wars. In the United States, anyone could enter and compete—until they lost. Both Marc Thorpe and Trey Roski were adamant about that. To them, robotic combat was a sport first, entertainment second.

In the United Kingdom, Tom Gutteridge and Steve Carsey turned the concept on its head. Mentorn fielded applications from fans of the show who wanted to build a robot and compete, but urged competitors *not* to build their robot until they heard that their application was being considered. Then, a few weeks before the taping of the show, they auditioned all the completed robots in a studio (and later, during the live events). A minority of those robots made it to the final taping. Preference was given to the best-looking robots and the most original designs. Starting with Series Two, Mentorn hired as host the chipmunk-cheeked Craig Charles, an actor on the BBC's hit science-fiction show *Red Dwarf*.

At U.S. robotic combat events, there were four weight classes, and one match followed another in quick succession. Matches were set randomly by software, with consideration given to ensuring that the frequency channels for each combatant didn't overlap. In England, there was only one weight class and there were long delays between bouts to adjust cameras and lighting. Matches were set up, according to Mentorn, "to ensure maximum drama, entertainment, and competition for viewers at home."

In England, the house robots were the stars, patrolling the *Robot Wars* arena, entering the action to dispose of one or both of the competitors at the end of a fight. Sir Killalot, Matilda, and the rest also constituted recurring characters for the young fans who wanted to see their favorite robots week after week. Though Marc Thorpe had included an orange-coned house robot in his first competition in 1994, Tom Gutteridge maintained the house robots were his own invention. "What drives audiences is familiarity," he said. "We needed regular characters who would be seen each week. We needed to have 'super robots' that the contestants could aspire to, and we needed robots that had a certain number of 'human' characteristics too, to drive the younger audience to the show."

In the United States, Thorpe developed robotic combat for a small group of animatronics-obsessed gearheads, and for the viewing enjoyment of a tech-savvy San Francisco audience. In England, Tom Gutteridge and Steve Carsey developed it with a mainstream television audience in mind. "I like to think this is the reason it's been successful," Gutteridge said. "The way it's shot. The way it looks. The house robots. Its more of a TV show, not the filming of a sport."

In the midst of the heated dispute between Profile Records and Marc Thorpe, Gutteridge monitored the Web forum and occasionally wrote the U.S. builders. In one post, he opined, "The TV format is very different from the live event for one very good reason. We don't believe (and neither did the BBC, or any of the other broadcasters around the world who have seen it) that the live format would sustain a long-running series on prime-time TV. We've proved that our format does, and it will probably go around the world."

¤

The U.K. *Robot Wars* did share one characteristic with its U.S. brethren: It changed the lives of a small group of tinkerers, garage inventors, and artists. In Poole, on the southern coast of England, Ian Lewis and Simon Scott saw the first season of the show and were smitten. The industrial designers had gone to high school together, to "a real nutters school, where students set fire to the desks," they said. They were sure they could do a better job than the original crop of contestants. "If you're going to make something and put thousands of hours into it, you might as well make it look good," Lewis declared. The result, the scorpionlike Razer, was one of the most impressive bots on either side of the Atlantic. A starter motor from an old Honda motorbike powered a hydraulic pump the friends bought at a junkyard for five pounds. The resulting weapon punctured opponents' armor with nine tons of force at its stainless-steel tip.

Hours before Razer was set to compete for the first time, Lewis and Scott realized their masterpiece was too heavy. To shed weight, they drilled dozens of holes into the five-millimeter-thick steel. When they finished, Razer was not only dangerous, but a beautiful kinetic sculpture.

The second season of *Robot Wars* was filmed in August 1998, just as the robotic combat landscape was freezing over in the United States. Razer passed the Gauntlet and Soccer rounds, then went head-to-head against a pyramid-shaped, chainsaw-armed bot called the Inquisitor. Razer started out ramming its rival and tearing holes in its armor, but then Lewis executed too sharp a turn, which sheared the drive motor and immobilized the robot. Like angels of death, the house robots emerged and dragged Razer to the flaming grate on the floor.

Mentorn gave Lewis and Scott the best-design award and made sure they were back the following year. This time, Razer looked even better. To allow Razer to self-right in the event it got flipped on its side, they added wings, which extended from the neck like the hood of a cobra. In the third season of *Robot Wars,* filmed in October '99, the game rounds were discarded for the more broadly appealing head-to-head combat. In its first match, Razer grabbed hold of a spike-wielding competitor called Aggrobot and started mercilessly tearing holes in its aluminum armor. Then disaster struck again, as the tiny valve that controlled the direction of the hydraulic fluid in Razer's crushing arm jammed in the "up" mode. The arm lifted up and stayed there, knocking the bot on its back wheels where it couldn't move.

Again, Mentorn made sure Razer stayed in front of TV viewers. They gave Lewis and Scott the best-design award for a second year and invited them and 15 other competitors to participate in a TV pilot for other European countries, dubbed the "First World Championships." For that competition, which Razer won in four straight battles, Gutteridge brought Marc Thorpe to the United Kingdom to present the awards. The Robot Wars creator spent most of his time soliciting Tom Gutteridge for ways to solve his legal standoff with Steve Plotnicki.

With *Robot Wars* watched by millions of people each week, Ian

Lewis and Simon Scott, along with the builders of a handful of the other most popular robots, became minor celebrities—signing autographs on the street, getting random phone calls from fans and e-mail requests for more information about building a robot. They asked Lewis's fiancée's brother, Vincent, to design a Razer website, and within months, it logged more than a 15,000 visitors.

By early 2000, Razer was one of the most recognizable bots on *Robot Wars.* That January, when Mentorn struck a toy deal with British toy manufacturer Logistix Kids, it announced that out of 100 robots, only Razer, Hypnodisc, and Chaos 2 would join the popular house robots on store shelves. For every dollar their robot toy earned, the builders would get a 10 percent cut of Mentorn's profits on the sale. It wasn't quite the 15 percent that Trey Roski was promising to competitors in the States. But for Scott and Lewis, at least, it looked as if *Robot Wars* was going to pay something.

While a few teams rode *Robot Wars* to fame and many others enjoyed it and competed vigorously, others sought a purer form of competition and were disappointed. Darren Brown and Stuart Reynolds were roboteers from the northern England town of Bradford. Brown was a computer technician while Reynolds worked at a wagon fabrication plant; both were avid competitors in other mechanical contests, such as motocross and bike racing. They came across the U.S. Robot Wars website back in the mid nineties and saw it as the ultimate mechanical sport.

Several years later, the first season appeared on local British TV, and they knew they had to build and enter a bot. But when the Series Two taping was announced in 1998, they were too busy building an animatronic dinosaur as an attraction for a local shopping mall. They jumped at the Series Three taping in October '99. They sent in their application and, without waiting for a response, started building their fighting robot.

It took nine months and cost about 3,500 pounds: Hefty, made of Kevlar and a glass fiber, reinforced polyester resin shell, had a pneumatic ramming spike on its front and a flowery paint job reminiscent of the Mystery Machine from *Scooby Doo.* At first, the robot weighed 82 kilograms—just over the 80-kilogram (176-pound) weight limit. Brown and Reynolds were forced to remove their self-righting mechanism, one of their cleverest pieces of home engineering.

When they got to the event taping at Elstree Studios in North London, however, they were surprised to find that other bots were over the weight limit and that the organizers didn't seem to care. Moreover, Brown and Reynolds found it difficult to keep Hefty in prime fighting condition: The pneumatic system needed to be refilled with CO_2 every two hours, but Mentorn asked the builders to queue up with their robots for the entire afternoon waiting for their matches. "The rules and regulations were not very well enforced," Brown said. "Mentorn didn't seem to care at all about ensuring that the competition was fair."

Sure enough, Hefty lost its first match against a hammer-wielding bot called X-Terminator, which drove right across the arena and planted its spike in Hefty's Lexan shell. When X-Terminator withdrew the spike, Hefty got hoisted up in the air, then fell onto its back. Without the self-righting mechanism, Hefty was powerless. "Pit! Pit! Pit!" the studio audience yelled.

Enter Sir Killalot. The massive house robot rolled over from its corner of the arena, picked up the hapless Hefty, and dropped it in the smoking pit. The TV audience vociferously approved. Brown and Reynolds left the taping disgruntled, their expensive robot destroyed. "When you're there and expecting a competition, it rankles a bit. They are making a TV program, and if you can accept that, it's fine," Brown said.

Other competitors, posting on the various Internet sites, also grew weary of Mentorn's *Robot Wars.* Rex Garrod was a 25-year veteran of the special-effects industry from Ipswich, England, and the maker of the popular Series Two flipping robot Cassius. Garrod railed against what he saw as Mentorn's feeble safety record. During the taping of Series Three, a robot ran amok after falling off a trolley and speared a stagehand through the ankle, putting him in the hospital for a day. Garrod subsequently pulled out of *Robot Wars,* citing safety concerns.

But he reserved his harshest invective for the house robots, whose size and might enabled them to beat up on the competitor robots. "I think that the damage done to our robots should be reimbursed, since it's for the sake of TV, not the competition," Garrod said. "It costs a lot of money to construct a good robot. If my robot was destroyed by a fellow competitor I would gladly shake his hand. But if my bot was wrecked by a minidigger disguised as a robot weighing over half a ton, I would be really pissed off."

Meanwhile, Hefty's and Cassius 2's demises were included in Series Three, which began airing on British television in December 1999. The series was watched by six million people, making it the second-highest-rated program on the BBC2. In case the message to TV executives back in the States wasn't clear, here it was again in the ratings mathematics they understood—robotic sports made for good television.

As 1999 progressed, Tom Gutteridge developed his own little problem to match Marc Thorpe's: He too was feuding with Steve Plotnicki, who believed that Mentorn's original license didn't grant it the right to exploit the Robot Wars name outside the United Kingdom.

Gutteridge argued that the value of the property lay in the format; the house robots, the style of the event, everything that Mentorn had created. The BBC, meanwhile, also stuck its nose into the feud, feeling it should be compensated for the house robots in any extension of the *Robot Wars* series. Negotiations that year between three sets of lawyers grew acrimonious and nearly sprouted into a whole new round of litigation. Plotnicki refused to let Mentorn's *Robot Wars* expand further without a new deal and threatened action over unlawful use of the Robot Wars name. "It was an unpleasant eight or nine months," Gutteridge recalled.

In the fall of '99, the BBC, Mentorn, and Profile struck a tripartite deal. Mentorn got a license to produce and distribute the show internationally, and the BBC shared in the revenue. "At the end of the day, we were on strong ground," Gutteridge said. "We held our nerve."

Plotnicki was less than happy with the deal. Back in 1999, he had wanted to stage his own competition in the United States and then sell the idea to a national TV network—and reap 100 percent of the profits. When the '99 Robot Wars got canceled, Plotnicki had to resort to plan B: allowing Mentorn's brand of robotic combat into the United States. "We had been trying to prevent that because we wanted to control it all ourselves," Plotnicki said.

In this case, Plotnicki's loss was Gutteridge's gain. With his new deal, Gutteridge was free to bring his show to the United States. The United States was the largest and most lucrative television market in the world. By extension, toys, supported by a popular TV show on American cable, would fly off shelves.

Mentorn had immediate interest from several networks in America. For instance, both the Discovery Channel and Viacom, parent company of MTV, were intrigued.

Comedy Central's new boss was also interested. Bill Hilary was the former commissioning editor of the BBC and the man who once oversaw *Robot Wars* for the British television network. Gutteridge claimed Hilary's first phone call in his new job in the United States was to him, to ask about bringing *Robot Wars* to the States. Accounts from the former friends differ on the result of that initial conversation. Gutteridge said he found a better offer elsewhere. Hilary said he didn't like the deal Gutteridge's representatives were offering, and that he didn't think Mentorn's game-show format would work in America. "I was much more interested in a real sporting event," he said.

In any event, Gutteridge also had interest from an independent producer in the United States named Bruce Nash, the creator of reality programming such as *World's Most Amazing Videos.* In 1999, a Nash employee sent Mentorn an e-mail, inquiring about licensing Mentorn's new *Robot Wars* sister program, called *Techno Games,* which featured remote-controlled robots in events like racing, rope-climbing, and swimming. Gutteridge said he wasn't interested in licensing the concept. By 2001, Nash would unveil his own robotic show in the United States for the Learning Channel, called *Robotica.*

Gutteridge narrowed his U.S. options down to the Discovery Channel and Viacom. He flew to Washington, D.C., to meet with executives of the Discovery Channel, while his agents met with executives at Viacom in New York. Gutteridge and his newly genial partner Steve Plotnicki weighed both offers and ultimately chose

Viacom, the third-largest media company in the world, which therefore was able to offer unique opportunities. Viacom programmers talked about airing *Robot Wars* on both MTV, with its heavy teen demographic, and Nickelodeon, which targeted the younger kids who would buy *Robot Wars* toys.

Just as Trey Roski and Lenny Stucker were preparing to shop the fledgling Battlebots to U.S. networks, Plotnicki and Gutteridge were plotting the return of Robot Wars to America. Competing versions of Marc Thorpe's idea were on a collision course.

Lenny Stucker and Trey Roski knew they had to act quickly. The Battlebots pay-per-view special had done poorly. It aired on the InDemand service in early 2000, on the night before the St. Louis Rams beat the Tennessee Titans in Super Bowl XXXIV. About 25,000 people paid $14.95 to watch.

But now the Battlebots team had a new asset in trying to sell their brand of robotic sport: a TV pilot, which would differentiate it from Mentorn's game-show-oriented fare. Stucker showed the pay-per-view tape to television programmers from L.A. to New York. He talked to the all-technology, all-the-time cable network ZDTV (now called TechTV), the Discovery Channel, ESPN, and the USA Network, which all showed varying degrees of interest.

He also had a promising conversation with the folks at the Sci-Fi Channel, owned by USA Networks. Sci-Fi president Bonnie Hammer had recently shelved her own robot project—dubbed *RoboDeath*—which she had originally conceived as a way to televise Mark Pauline's SRL performances. Hammer had approached the uncompromising SRL chief, who agreed to listen to proposals but made it clear he wasn't going to change anything about his art. Hammer experimented with several ideas, including the unlikely one of teaming Pauline with her old mentor, Vince McMahon, impresario of the World Wrestling Federation. Pauline actually flew out to Stamford, Connecticut, to meet McMahon, and called him a "nice guy who seemed pretty excited by the concept." One of the original proposals involved staging SRL shows, with McMahon or one of his wrestlers acting as a play-by-play announcer, slipping in and out of the carnage in a shielded vehicle.

The concept died when McMahon took the WWF from USA Networks to Viacom. Later incarnations of *RoboDeath* proved too expensive to produce. Bonnie Hammer's people got in line to speak with Battlebots.

But Roski and Stucker ultimately settled on Comedy Central. Bill Hilary, the former BBC commissioning editor, said that Battlebots was more up his alley than *Robot Wars.* The 10-year-old comedy network's ratings were flat. An actual sports program was something different to attract viewers and jolt the numbers.

After weeks of negotiations, Lenny Stucker and Trey Roski hammered out a somewhat awkward partnership with Comedy Central. Roski and Battlebots would stage the event. Then a Comedy Central–affiliated production company, called First Television, would take the footage and edit it into a TV show, inserting the commentary of its own TV personalities.

Compromises were made by both sides. Comedy Central wanted to replace the tuxedoed ring crier Mark Beiro with a stand-up comic, but Stucker felt that the professional boxing announcer added an element of sports gravitas, and insisted he stay. Comedy Central wanted stand-up comics and buxom models interviewing the builders after each fight. Roski and Munson felt that was cheapening their sport—but gave in. Comedy Central also wanted to limit the number of builders who could compete. Trey Roski maintained that anyone who showed up at a Battlebots event should be able to enter.

The first Comedy Central Battlebots event was held in June 2000, and took place at the birthplace of robotic combat, Fort Mason in San Francisco. Builders brought 75 robots to the competition and this time, they faced the trappings of a full-fledged TV production. Cameras turned the Battlebox into a fishbowl. Twin comedians Jason and Randy Sklar, and blond *Baywatch* actress Donna D'Errico, occupied the staging area outside the Battlebox, playfully needling competitors after their matches. Bill Nye "the Science Guy" conducted extended interviews with the competitors and offered a pseudotechnical analysis of the most interesting robots.

There was much for him to analyze; builders were upgrading

their robots with each competition. Biohazard, Vlad the Impaler, Rhino, and SLAM were all improved and ready to fight. In addition to his super heavyweight Minion, Christian Carlberg brought a ramp-shaped heavyweight named Overkill, which sported two enormous racing wheels and a giant blade, which slammed back and forth as the robot reversed direction. Carlberg, an enthusiast since his first competition in 1997, was the first builder to bring more than one robot to the tournament in order to maximize his TV exposure.

Will Wright and his 14-year-old daughter, Cassidy, built a bladed middleweight named Chiabot. A fake plant sat on the robot's base, and a 10-pound "annoyance bot" named "Chia's Little Helper" emerged from the mother ship during battles to jam the opponent's wheels (though this never quite worked).

A new Mauler 2000, refurbished by Supreme Commander Tilford, had about five times the power of the old version. Unlike Blendo or Ziggo, only the very top of Mauler rotated, violently flinging three maces around as it spun. General Henry adorned the bot with decorative swirls of masking tape and magic marker. His older brother, General Morgan, sporting a wild, red Mohawk, drove the robot in battle. With his punk guise and keyed-up antics, Morgan was the perfect foil for the Comedy Central television cameras.

In its first match—shown in the first episode to be broadcast on cable television—the new Mauler went up against Jim Smentkowski's Nightmare. The battle lasted only 40 seconds. Mauler's maces blasted one of Nightmare's back tires, shattering the wheel hub. A few more hits and Nightmare was finished.

A few British bots crossed the Atlantic once again. Former formula race car mechanic and Web programmer John Reid brought his Lexan-sheathed hammerbot, Killerhurtz. This was Reid's third Battlebots tournament, and he was beginning to notice that the sport was evolving differently here in the States. Roski was promising the builders a $1,700 royalty each time their robot appeared in an episode on TV, and the prize for winning the heavyweight nut was $2,000 and would later climb to $8,000. In the *Robot Wars* competitions Reid had attended for Series One through Three, there were no appearance fees and no prizes. Tom Gutteridge said the BBC wasn't paying enough to make it possible. "It seemed Mentorn was falling behind," Reid said. He and other British competitors participating in Battlebots would take that observation back to England.

After the competition, Comedy Central sent producers and crews to many of the veterans' homes, to do short segments on the builders themselves. Producers poked TV cameras into tool-cluttered garages and basements and interviewed the builders and their long-ignored spouses. It was the hardest part about making the new show, recalled *Battlebots* writer Dave Rygaliski. "The builders were unsure and defensive. Most had never been on TV before. Frankly, aside from Stephen Felk, very few had outgoing personalities."

Felk loved the media attention: Voltarc defeated Biohazard that season, maneuvering underneath the low-rider's titanium skirt and hoisting it into the air. No one had ever so decisively gotten the advantage of the former champ. Though he lost to Gage Cauchois's Vlad the Impaler in the heavyweight finals, when Comedy Central arrived at his home, Felk was hamming it up, ushering the camera crews into his impossibly cluttered living room, where he declared, "I've basically had no life for four years, but I have a pretty decent robot."

Amplifying their militaristic shtick, the Tilfords did well in the public-relations department too. From his home in Portola Valley, wearing his favorite straw hat, the Supreme Commander said, "I won't stop until I've achieved total robot domination worldwide."

The Mohawked Morgan advised: "You've got to make the other robot your bitch."

"Calling all American Robot Warriors," said the press release. "If you've got a battle-ready robot and you're itching for a fight, you can't afford to miss this once-in-a-lifetime opportunity.

"What: The U.S. Robot Wars heavyweight champion.

"How much: a $5,000 appearance fee for all competitors.

"When: July 22 to 23, 2000, in London, England. All freight and flight arrangements will be made for you. It couldn't be easier."

The last time Tom Gutteridge had appealed to the U.S. robot makers, in 1998, the community had responded with nasty invective about Mentorn's relationship to Profile. Now two years had passed, Gutteridge had a new deal with Plotnicki to expand Robot Wars all over the world, and he knew how to appeal to the builders' hearts.

"A free trip to England!" Morgan Tilford exclaimed. While his dad, the Supreme Commander, wouldn't be able to attend, Morgan and his brother Henry didn't want to pass up the opportunity. A show on MTV would be great for the sport, but it could be bad for Battlebots, so the Tilfords called Trey Roski that summer to make sure going to London was okay. Roski didn't want them to go, but said he wouldn't hold it against them if they did. He also told them to scrutinize any contracts they signed. With the possibility of Battlebots toys on the horizon, he didn't want the team signing away the image rights to Mauler 2000.

The Supreme Commander had a concern of a different sort: 22-year-old Morgan, a student at Foothill Community College, had three final exams the week after the event was taped. Tilford made the travel coordinators at Mentorn promise they would get Morgan and Henry back to California on a direct flight immediately after the tournament. Tilford also sent his friend Alan Vermette, an electrical engineer, with his sons and the spinning robot.

Seven American teams made the trip. Patrick Campbell signed up with his hammerbot Frenzy. New Jersey–based Andrew Lindsay went with his bot, Spike, which carried on its frame the disembodied head of a baby doll. Dan Danknick flew over and met British builder Adam Clark, who had built the dual-wheeled robot Mangulator. Team Ram Tech came from Florida with their trapezoidal Ramstein. Todd Mendenhall, the head of SORC, brought his partner's robot, Meolnar. Most were attracted by the free trip and the large purse. "Five thousand goes a long way in our shop," Mendenhall said.

Steve Plotnicki was also there, at the St. Albans studio outside London. He was helping his Mentorn partners craft the return of Robot Wars to America, but he kept his distance from the U.S. builders.

For three of the teams, the competition quickly proved a bitter disappointment. There were supposed to be eight squads in all, but one canceled at the last moment. The competitors were told only four robots would be able to enter the competition—and only those four teams would get the appearance fee. Then each robot and its driver was auditioned for maximum appeal to the MTV audience.

Team Mauler was an obvious choice. Morgan's giant foam cowboy hat and manic energy made a perfect fit. The destructive Frenzy, doll-headed Spike, and Dan Danknick's Mangulator were also picked; the others were left with nothing to do but watch. Todd Mendenhall, who had traveled from L.A. for $5,000, was apoplectic.

The filming went quickly: Mauler beat Andrew Lindsay's Spike, ripping the doll head off and tossing it across the arena in a shower of sparks, while Frenzy decimated the lethargic Mangulator. The two victorious robots then faced off in the final bout, and Frenzy's spike broke loose while hammering at Mauler's spinning lid, leaving Patrick Campbell's bot vulnerable. After two fights, the Tilfords and Mauler were the master of the new U.S. *Robot Wars.*

But the competitors remember their U.K. trip more for what happened next than the taping itself. Working with a limited budget, Mentorn had negotiated a bulk travel discount with Continental Airlines. The Tilford boys and Alan Vermette took a shuttle that night from London to Glasgow, Scotland, where they were due to catch a transatlantic flight back to San Jose. When they got to Glasgow at 10:00 P.M., to their surprise, the airport was shutting down for the day. They got in touch with the Mentorn travel coordinators, who informed them their itineraries were wrong. They were booked on a flight to Newark—in two days. They had no place to stay, and Morgan was going to miss some of his exams.

They found a hotel room in the gritty airport neighborhood of Scotland's capital and paid for the room using the MTV credit-card number, which they took from a receipt for a pair of black pants that the producers had bought Morgan for the show. But their sense of satisfaction quickly passed: An air-conditioner dripping water in their rooms kept them awake until a maintenance man could be summoned.

Henry Tilford didn't notice anything unusual about his older brother's behavior during the next two long days. "The only thing that was strange," he said, "is that while me and Alan were exhausted and slowing down from the competition, Morgan was staying alert, worrying about how he was going to get home for his exams."

Though they were sharing a room, Henry said he was so tired that he didn't notice that his brother wasn't sleeping.

The next day, the robo-warriors toured Glasgow. The following morning, they caught their flight to Newark, and then to California. Later, they would learn that other teams also suffered travel snafus. By offering a free trip and a chunk of cash, the producers had enticed the robot builders into helping them make a TV pilot. But either Mentorn or Continental hadn't given proper attention to the task of returning them home.

By the time Henry and Morgan got back to the Bay Area, Morgan hadn't slept in a week and was speaking in disjointed phrases. He had already missed one exam and tried to take another, an oral test for a course in public speaking. The topic of his speech was robotic combat. "I would have gotten an A," he said later, "but I spoke for too long and scared everyone."

The following day, the Supreme Commander told his son he was taking him to the hospital. Paranoid and insistent about taking his last exam, Morgan made him promise that he wouldn't be forced to take any medication. His father promised, then took him to a doctor in Palo Alto who immediately diagnosed him with bipolar disorder, a serious mental illness that often manifests itself during a stressful event. The doctor tried to inject him with a psychotropic drug to quell the manic episode, but Morgan escaped and fled the doctor's office on foot. He was found later that day, lying inert in a line of hedges near the Stanford Park Hotel, and was committed to the inpatient psychiatric ward of Stanford Hospital.

The British roboteers met that fall at the Oxford home of John Reid, builder of the heavyweight hammerbots Killerhurtz and Terrorhurtz. Reid, along with the other British competitors who had attended both U.S. and U.K. competitions, was worried. *Battlebots* at least attempted to compensate participants, but Mentorn seemed oblivious: Not only were there no appearance fees, but the presence of the heavily armed house robots ensured that competitors' robots would be destroyed. Even worse, sponsorships weren't allowed on the BBC, because the network considered it not a true sport but "sports entertainment," a category that included professional wrestling, so the builders had even less recompense to make up their investment. "Mentorn wanted it both ways," said Ian Watts, builder of Big Brother. "They were treating the show as a drama instead of a sport, but they weren't paying the roboteers as actors."

The British builders also hated the contracts they were forced to sign at *Robot Wars* and its spinoff, *Techno Games.* Clauses reserved for Mentorn the exclusive merchandising rights and instructed the builders not to enter their robot in any other live or televised event "deemed by us to be in competition with *Robot Wars.*" *Battlebots* used similar language, but Trey Roski at least was paying for those exclusive rights. For the British builders who participated in both *Robot Wars* and *Battlebots,* the possibility of toys made the situation even stickier. Which organization should they sign a toy deal with? That fall, said Ian Watts, "You would have been hard-pressed to find any roboteers happy with Mentorn."

The builders met at Reid's house, and then took their meeting to a nearby pub. They decided to send a letter to Tom Gutteridge listing their grievances. Two weeks later, before the letter was even sent, Gutteridge called a meeting of 32 veteran teams at the Mentorn offices in London. No one thought it was a coincidence.

The meeting was held at the nearby Charlotte Street Hotel in London. Tom Gutteridge stood up and gave a rousing speech. "We wouldn't be here without the roboteer community," he said, and promised that the offending clauses in the contracts were mistakes. The roboteers were free to compete in other tournaments, he said, although the robots would have to be at least superficially different (for instance, a new paint job and altered name). If they chose, the roboteers could sign a toy deal with *Battlebots,* although that would preclude a deal in the United Kingdom. Gutteridge also promised to introduce prizes of up to 2,500 pounds (around $3,600 at the time) and appearance fees of 100 pounds (roughly $140) for all comers in the next *Robot Wars* series. The half-dozen teams chosen to have their robots made into toys would receive a 2,000-pound (then in the neighborhood of $3,000) advance on their royalties.

He couldn't do better than that, he said, because the BBC didn't pay

Mentorn enough to allow it. Left unsaid was the fact that Mentorn also was funneling another piece of the proceeds to the Robot Wars LLC, aka Steve Plotnicki, the owner of the Robot Wars trademark.

The builders were mollified, at least temporarily. "Tom allayed our fears," said Ian Watts. John Reid added, "He convinced us that all these issues were more a matter of oversight than deliberate." Both Watts and Reid would continue to participate in both events, but they would sign toy contracts with *Battlebots.*

A year later, Tom Gutteridge would insist that the fall 2000 meeting between the builders and the show organizers was routine, and that the roboteers had simply "misunderstood what their rights were." The British roboteers, however, told the story differently. Even though they would continue to complain bitterly about Mentorn's stinginess, in their version, they had banded together and won a small share in the bounty that was the success of *Robot Wars.*

With his son in the hospital that fall, the Supreme Commander fired off a profanity-laced letter to Mentorn. On the Delphi *Battlebots* forum, he informed the U.S. robot community about Morgan's breakdown, which had been triggered, at least in part, by travel snafus after the MTV *Robot Wars* pilot. The builders responded with sympathy and affection. "All our best to your son, Tilf," wrote Gary Cline. "I have two of my own and I know how you must feel. There's nothing worse. You're a good dad."

Stephen Felk wrote, "Morgan, feel better quick, Donna D'Errico misses you!"

"Nothing that we do in this hobby or sport is worth the health of a single human," Todd Mendenhall posted. "Let's keep this in mind as we go forward."

Trey Roski and Greg Munson visited Morgan in the hospital. They brought along a videocassette—the inaugural episode of the new Comedy Central show, *Battlebots.* The show featured Mauler's victory over Nightmare, and the segment on the Tilford family where Morgan proclaimed, "You've got to make the other robot your bitch."

The cousins met Morgan in the TV room of the psychiatric ward, but before they could play the tape, another patient grabbed it. "Wait, wait!" he cried out. "Maybe we shouldn't watch that. Let's take a vote. Who wants to watch it? Let's all sing instead."

Another patient was rocking back and forth on the couch, nearly catatonic, imploring, "Just decide, just decide, just decide."

After a brief chase, the cousins recovered the tape and played it on the ward VCR. Morgan enjoyed it. He was released nine days later, on a regimen of the drug lithium to control his disorder. Of his fellow robot builders, he said, "They still respected me. That was the most surprising thing."

The Tilfords quickly went to work. They built a new Mauler, more powerful than ever. Its lid spun at 215 miles per hour, rotating two steel maces in a dangerous blur. It was called Mauler 51-50, a reference to the California code stating that any person can be committed who, "as a result of a mental disorder, is a danger to others, or to himself or herself."

A few weeks later, Morgan was back in the spotlight, doing his eccentric Mauler shtick on—of all places—*The Tonight Show with Jay Leno,* because when *Battlebots* premiered, it reached a level of national recognition that no one could have predicted. "It was and continues to be a surreal mix of medical situations and celebrity," said his father.

Battlebots debuted on August 30, 2000, in the 10:30 P.M. slot, right after *South Park,* and 2.1 million people tuned in, making it the highest-rated premier ever on the cable network.

For the builders, it was thrilling, it was what they had been anticipating for all those years, it was . . . a little disconcerting. This was Comedy Central, after all, and there was an obvious lacing of parody, which at times bordered on crude, frat-house humor. Cohost Randy Sklar asked the San Diego schoolteacher Dan Rupert, maker of the super heavyweight Grendel, "Do you feel like the winner of this battle will have his pick of women?" After Rupert's other bot, Alien Gladiator, lost to Grant Imahara's Deadblow—which had lost its hammer during the fight—Donna D'Errico said to one of Rupert's meek teenage students, "It must be tough losing to a guy who lost his pecker." And after another fight, D'Errico complimented Vlad the Impaler maker Gage Cauchois, "I was noticing you have really good shaft control." He didn't seem to get it.

Many of the show's elements seemed designed to wink at the audience, even as the competitors were intently preparing and fighting their robots. The ringside interviewers, D'Errico and the Sklar brothers, for instance, knew nothing of the gearhead sensibility and probably didn't care about it either. They were as much real sports journalists, the joke went, as the robot builders were real athletes.

For the builders who had taken the sport so seriously for so many years, it was unsettling. "The representation of the sport is demeaning to the Battlebots staff and more importantly, to the builders," one robot builder posted on the Web forum, articulating the majority sentiment in the community.

Still, for many in the robotic combat world, the undercurrent of parody was nothing but a nagging irritation. After all, the lunacy of the whole endeavor was a part of its appeal, as Debbie Liebling, the senior VP of programming at Comedy Central pointed out. "These are inanimate objects with no nerve endings," she said. "No one gets hurt. It's a safe environment and it's playful. . . . It's a lark really."

Moreover, not only were the robot makers on television, being interviewed and having their fights shown, they were also getting paid. Finally, after years of putting their money and time onto a one-way conveyor belt, there was reward. A royalty check in the amount of $1,673 was sent to builders each time their robot appeared on the show.

In the fall of 2000, as *Battlebots* played on Comedy Central, the rest of the media started to pay attention as well. Nearly every paper and magazine weighed in on the unexpected TV phenomenon. *The Los Angeles Times* wrote about Minion creator Christian Carlberg, then did another article in their Sunday magazine on Trey Roski and his powerful father. *The Washington Post* ran a story featuring Supreme Commander Tilford, depicting him horsing around in his garage workshop and singing boozy robot songs with Morgan and Henry. *The Toronto Sun* focused on Canadian Derek Young, maker of the middleweight walker Pressure Drop. *USA Today* wrote, "Since its debut, the techie transmutation of cockfights and tractor pulls has become the second most popular program on Comedy Central."

Few of the articles mentioned Marc Thorpe. Throughout the heady success of the first two years of *Battlebots,* Roski and Greg Munson tiptoed around the real origin of their sport. The truth was still a legal minefield.

After its first successful season, Comedy Central approved the production of another 18 episodes and *Battlebots* planned another event that fall in Las Vegas. Again the Battlebox was deconstructed and shipped to the All American SportsPark in Las Vegas. This time, the hazards were even more deadly, with the steel hammers called Pulverizers at each corner of the arena. Again, there were more and mightier robots than ever, 126 in all.

Alexander Rose and Reason Bradley from Sausalito retired their pneumatic puncher, Rhino, and in the two weeks before the competition, built the far more deadly Toro. It weighed 340 pounds and carried a set of bullhorns on top of its curved flipper. A 1,500 PSI pneumatic cannon powered the flipper, which exerted 7,000 pounds of force and was capable of flinging other super heavyweights five feet in the air. Toro lost its third match to the obviously wedge-shaped Atomic Wedgie but owned the super heavyweight rumble, flipping all the other robots in a quick three minutes.

Mauler 51-50 lived up to its name by destroying itself in appropriate fashion. Fighting Ian Watts's flipping heavyweight Big Brother, Mauler lost one of its weapons, began convulsing from its own unbalanced gyroscopic forces, and flipped itself upside down. Smoke poured from its base. Such self-destruction would become known among the builders as "the Mauler dance."

Carlo Bertocchini, finally, pulled himself back to the top of his weight class. After losing one year ago to Vlad the Impaler, and then last July to Stephen Felk's Voltarc, he had refitted Biohazard with stronger titanium skirts, a new lifting arm, and refurbished electronics. Bertocchini scored three straight knockouts in the Vegas tournament and defeated Vlad the Impaler in the championship, 27 to 18.

There were new competitors too, attracted by the growing reputation of the sport. Michael "Fuzzy" Mauldin caught the first season of *Battlebots* on Comedy Central. He was no mere garage gearhead: As a Carnegie Mellon professor and the creator of the Lycos search engine in 1994, Mauldin rode the Internet boom and cashed out at the right time. By 1997, he was a millionaire retiree and the father of three young children. "Being a software type, I thought that building a machine in the real world would be a new and fun challenge," he said.

To clear the time and the space to build the orange, rectangular plow FrostBite, Mauldin stopped flying his stunt plane and moved his Porsche 911 outside the garage at his Irwin, Pennsylvania, home. Carried along by his enthusiasm for the new sport, he ended up buying out the entire stock of utility motors from the Minnesota-based National Power Chair, inadvertently driving the prices up, leaving the other bot builders temporarily disgruntled. But they would have to get used to it: As their sport increased in popularity, the long-time builders were going to be contending with a whole new class of competitor.

Emboldened by the success of the first season, Comedy Central continued to walk the fine line between sport and parody, but the builders were still wary. They loved the coverage of the fights, as most of the matches were shown unembellished, overlaid only with the energetic commentary of hosts Bil Dwyer and Sean Salisbury.

A few of the comedic segments between matches were particularly well-regarded. One involved Carlo Bertocchini's cat, Elvis, and earnestly reported that the cat had lost the use of his hind legs after getting hit by a car. The resourceful Bertocchini, the segment claimed, had built a robotic walker that the cat controlled with its front paws. The clip featured the Biohazard maker and his wife, Carol, staring proudly as their paraplegic pet roamed around the garage. Of course, the whole thing was a gag—Bertocchini was controlling the cat-walker via remote control. He was flooded with letters and e-mail from pet lovers around the country anyway.

But other Comedy Central gags laughed at the robot builders, not with them. The most conspicuous offense that November in Las Vegas was inflicted on Mark Setrakian. The special-effects builder was truly the Michael Jordan of the sport, constantly raising the bar technologically and aesthetically. His Mechadon and Snake could be museum pieces. After the Snake lost its second match, getting gored at its defenseless center by Dan Danknick's wedge-shaped War Machine, Setrakian walked out of the Battlebox and was interviewed by the second-season "botbabe," Heidi Mark. The Playboy model motioned to the 13-foot-long mechanical python.

"Would you say you build robots as an extension of yourself?" she asked, stuttering over the line the Comedy Central writers had handed her.

"Nice try," responded Setrakian, getting the gist anyway.

"I had to ask. And if you did, I'd like to get to know you better."

Setrakian arched his eyebrows in disdain, and the rest of the builders just shook their heads. Anyway, it was difficult to get too upset, because the crest of attention and media exposure kept getting higher.

Trey Roski's goal back in 1995 had been to get robotic combat on *The Late Show with David Letterman.* Now it was finally going to happen—but on *The Tonight Show with Jay Leno.* That fall, Comedy Central writer Dave Rygaliski sent a tape of the first few *Battlebots* episodes to his former employers at NBC. They loved it and devised an elaborate, extended feature on *Battlebots.* Two NBC special-effects workers were commissioned to build Leno his own remote-controlled combat robot, ChinKilla. Built in two weeks, it weighed 400 pounds and sported thick steel armor and two spinning blades on each side of a trucklike frame. A large, polycarbonate Jay Leno face featured a pneumatically powered chin that violently popped outward.

In the fall of 2000, *The Tonight Show* ran *Battlebots* bits on four separate episodes. Roski and Greg Munson made sure the new sport put its youngest faces forward. Thirteen-year-old Lisa Winter and 30-year-old Christian Carlberg appeared as guests on the first segment, sparring with their bots outside the NBC studio. Leno watched and theatrically vowed to build his own robot.

In the second segment a few days later, Leno introduced ChinKilla to his audience. Then Scott LaValley of DooAll, Donald Hutson of Diesector, and General Morgan of Mauler gathered to demonstrate their bots against set pieces like a pile of pumpkins and a stack of bricks. Viewers would get to vote on the NBC website for who they wanted to face ChinKilla in battle. First though, Leno quizzed them all about whether they had a girlfriend. Inexplicably, Morgan declared, "Chicks dig *Battlebots,* especially German chicks, I don't know why."

Afterward, LaValley and Tilford went to a nearby bar with Trey Roski and Greg Munson, who had been watching from the studio audience. They made the bartender turn down the music when the Leno segment came on the bar television. Afterward, the foursome wandered over to the strip club across the street, where a friend of Roski's worked, and duly celebrated the great media exposure.

Morgan's eccentricity apparently intrigued the voting audience, so in the third *Tonight Show* segment, Morgan and Mauler fought ChinKilla in a one-fourth-scale Battlebox built by NBC. Mauler beat ChinKilla twice, but the TV audience saw faked footage of the Tilfords' spinner getting flipped by the NBC robot.

For the final bit, Leno flew to Vegas for the actual event. Before his special-effects team fought ChinKilla against Trey Roski's souped-up Ginsu in a demonstration match, Leno roamed the arena, grilling spectators about their interest level and teasing participants about their sex life. He encountered the Tilfords and asked if they abstained from sex the day before the competitions. "The robot abstains from sex," the Supreme Commander shot back.

The New York Times applauded the extended segment. "Mr. Leno's stunt illustrates how perfectly this mechanized meta-sport merges traditional television's celebration of violent sport—boxing, wrestling, football—with today's ironic sensibility, which has become the defining characteristic of Comedy Central."

Eight million people watched Jay Leno's program each night in the fall of 2000. Suddenly, *Battlebots* was a hot TV show, and they were all becoming stars. Carlo Bertocchini was signing autographs. ("It's not the kind of thing a mechanical engineer is used to.") Stephen Felk got recognized on the street in front of his Nob Hill apartment building. ("Hey battledude!") Young Lisa Winter received e-mail from admirers. ("Lisa, your robot is hot and so are you.") And Jim Smentkowksi's website, robotcombat.com, was drawing 15,000 readers every month. After any episode that featured his bot Nightmare or his similarly designed lightweight, Backlash, it spiked even higher.

For the robot builders, the customary 15 minutes of fame ticked by, and afterward, to everyone's surprise, the camera lights still shone brightly on the world of robotic combat.

Trey Roski would have been happy if the story had ended right there, at the end of the year 2000. *Battlebots* was an unexpected hit; to the builders, he was a hero, the guy who saved the sport they loved; his powerful father, who hadn't seen a way to make money in competitive robotics, was being proven wrong; toy companies were interested in licensing Battlebots designs; Steve Plotnicki had dropped his lawsuit against Battlebots in New York, though he continued to contest Marc Thorpe's bankruptcy in Santa Rosa.

Meanwhile, MTV had passed on its *Robot Wars* pilot, which had been filmed at such great personal cost to the Tilford family. The property was now floating unclaimed in the Viacom development pipeline, presumably until the option ran out and it reverted to Tom Gutteridge and Plotnicki. *Battlebots* had the American market to itself.

It might have stayed that way too, if not for the philanthropists behind the robot competition FIRST.

Chapter 8

The Square of 2.70

I'm not an educator. I'm not a sociologist. I got a little company. I invent things. I like to look at the world like everybody else looks at it, but when I see something new and identify a problem, I try to do something about it.

—Inventor Dean Kamen

Most of the robots took forever to crawl to the end of the table. Some failed to move in a straight line, then hurled themselves off the side, smashing into little pieces on the floor if no one caught them in time. A few just sat there forlornly, while others took off too quickly and didn't brake, banging into the wooden wall past the finish line and splashing their cups of water all over the playing field. Only half of the students' machines properly traversed the eight-foot-long course.

George Lechter was 18 that year, 1973, a sophomore from Cali, Colombia. He was the son of a heart surgeon who wanted to go to Harvard after seeing the 1970 film *Love Story,* but ended up at MIT instead. He didn't know much English at the time, but he knew enough to realize that being a student at the Massachusetts Institute of Technology in the early seventies was unequivocally uncool. "Bill Gates made the nerds fashionable," Lechter said. "But in those days, we were pathetic. Girls got on the bus and went right past the MIT campus. They were allergic to us."

At first, the contest that capped Engineering Synthesis and Design seemed frivolous and annoying: The class, referred to among the students by its course book code, 2.70, was already deadly dull. Students endured hours in a stuffy room and walked out of each lecture with a different equation to master. The professors told Lechter and his classmates that their performance in a design competition would constitute one-third of their final grade. Each student received a kit of random parts: They would have to use them to build a machine capable of transporting a paper cup of water across an eight-foot-long table without spilling a drop.

Few of the students knew anything about building an actual working device. "We thought they were full of it and resented that a large chunk of our grade was going to come from midnight lathe experiences," Lechter said. "I don't think we expressed these feelings to the professors, but amongst the students, we kind of felt the whole thing was silly."

In the kit each student received was a hodgepodge of random parts: a small DC electric motor from a Polaroid camera, a rubber band, a few pieces of pine board, some tongue depressors, a welding rod, piano wire, and a one-pound sack of sand. Students could only use those components. Making it work would be a nightmare.

Then the students started building. They wandered into the department's machine shop and spent three days before the contest thinking about little else. The competitive juices started to flow: Whose design intellect and inventing skills would prove superior? Excitement began to build as rumors of the most formidable machines rippled through the class. The night before the race, most of the participants didn't sleep. Every minute went into perfecting their creations.

George Lechter's machine was eight inches long, with a drive wheel at the front connected to the DC motor and two wheels at the back. The cup of water was suspended like a pendulum from the piano wire, which was bent into a semicircle. The vehicle's frame was cut from the tongue depressors. Lechter called his contraption "O' Angelica"—after a whorehouse in Cali that occupied a special place in his imagination.

While other machines expired midcourse, or wandered off the table, O' Angelica completed the race in 2.495 seconds without spilling any water. As the evening competition wound down, the time proved unsurpassed; Lechter had won. The next day, a front-page article in the campus newspaper, *The Tech,* showed the young Colombian bent proudly over O'Angelica.

Thirty years later, as a high-tech entrepreneur living in Miami, George Lechter remembered the competition fondly. "You really had to think creatively, without an example to guide you," he said. "Putting things together from different angles is what 2.70 made us do."

Long before there was robotic combat, there were more peaceful forms of machine contests: *task-based robot competitions.* Instead of one-on-one fights to the death between remote-controlled machines, these robots vied to outperform each other in specific challenges, such as elevating a Ping-Pong ball into the air, or herding Ping-Pong balls into a plastic trough. These early games taught college students about inventing from scratch, using machine tools, and how to abide by the unforgiving laws of physics. And the grandfather of all task-based-competitions was the 2.70 design contest, held each year at the Massachusetts Institute of Technology.

The first competition took place in 1970, while Mark Pauline was still horrifying teachers at Eckerd College in Florida and Marc Thorpe was crawling around the basement of the UC-Davis art building with his pants around his ankles. A team of several instructors in the design and control division of the mechanical-engineering department of MIT, led by future Texas A&M University chancellor Herb Richardson, gave 50 sophomores a plastic bag full of random parts. The challenge was open-ended that first year: "We asked them to design a device that would perform some function, or be a new concept," Richardson said. "It was very unstructured." Unfortunately, antiwar riots shut the campus down for two weeks, curtailing the inaugural competition.

Richardson's challenge was part of a pedagogic revolution within the mechanical-engineering department. World War II and the race to build atomic weapons had firmly rooted the engineering fields in the disciplines of math and physics. Students solved problems by plugging numbers into complex differential equations, or by working on the first generation of massive, room-size computers. When students were, on occasion, asked to demonstrate practical knowledge—for example, to sketch a three-speed gearbox—they often just headed back to their fraternity houses or dormitories and copied the designs from 20 years' worth of previous students' work. The disciplines of the garage inventor were fading away.

Bob Mann, one of Herb Richardson's colleagues, found this troubling. In the fifties, working for companies like Raytheon, Mann designed power supplies for the first generation of air-to-air missiles. He later worked on the biomechanical prosthesis called the "Boston Arm," which could detect brain signals and move in place of an amputated limb.

Throughout the sixties, as a professor at MIT, Mann campaigned to reconnect the theory of mechanical-engineering science with the practical tools and skills of the gearhead. He wanted to get students back into the metal shop, using such tools as drill presses and lathes and getting their hands dirty while wrestling with the demons of design and creativity. "My mantra was, there is no way to learn how to design but to do it," Mann said.

Herb Richardson staged his design competition in 1970 as part of Mann's efforts, and did it again in 1971. That year, Richardson and his staff of assistant professors and graduate students concentrated on improving the challenge. Instead of an open-ended design contest, they decided the machines should perform some specific task. They settled for a simple race. So, one by one, the student's creations lumbered down a sloping ramp, powered by the energy from a single rubber band and using sand as a ballast.

By 1973, the year George Lechter won, the competition was becoming an MIT tradition. The professor in charge that year was a British-born mechanical engineer named David Gordon Wilson, who traveled the MIT campus in a recumbent bicycle of his own invention and had an outspoken aversion to cigarette smoke.

In a photograph accompanying the article reporting George Lechter's victory, straggly-haired hippie students can be seen lined up four or five deep, leaning over each other to catch a glimpse of the action. The 2.70 competition was turning into something of a spectator sport at MIT.

But the professor who would take it there—who would fully convert the 2.70 contest into a campuswide, and then truly a national phenomenon—was one of Bob Mann's graduate students. He had helped administer the design competition since its inception, and he took over the class from David Gordon Wilson in 1974. His name was Woodie Flowers.

Flowers was a southern gentleman in the full sense of the phrase—a tall, lean, occasionally mustached showman with a predilection for bow ties and an almost empathic connection to his students.

He was born in 1943, in the backwater, racially polarized town of Jena, Louisiana, about 140 miles northwest of Baton Rouge. His father, Abe Flowers, the owner and sole proprietor of Abe's Welding Shop, was known around town for fashioning bizarre mechanical contraptions that defied belief.

One Abe Flowers invention was called the Slab Kicker, and solved the problem of how to safely transport heavy pieces of scrap wood from portable sawmills. These were common in Louisiana at the time, producing the crossties for the railroad beds that were stitching together the nation in the early twentieth century. Workers fed eight-foot-long tree trunks into the portable mills, and with four cuts, the mills produced an eight-foot-by-eight-inch-by-eight-inch railroad tie, plus four "slabs" that were discarded. A few poor souls were hired to drag the slabs to a huge bonfire, and they would scald their faces and hands as they tossed the wood into the inferno.

Abe Flowers had a better solution. The Slab Kicker sported triangular teeth spinning at 2,000 rotations per minute. It launched the slabs like field goals, arcing them into the air and down into the fire. Woodie Flowers remembered it as "an inside-out food processor on steroids. Dad could make every slab of wood hit the fire."

An Eagle Scout and butterfly collector, the younger Flowers joined the fraternity of mechanical obsessives at age 14, when his uncle got him a 1946 Dodge four-door sedan. He carefully guided it home and did what any self-respecting youngster at the time would do: He made a hot-rod out of it. He discarded the body, removed the front suspension, and replaced it with an antique truck beam axle. He built the new body from scrounged parts (including a 1922 vintage front fender) and handcrafted the body panels himself. And he replaced the clunky Dodge engine with a totally rebuilt Oldsmobile V-8. "In the end, the only parts of the original vehicle left were the back axle, the shortened frame and the radiator," Flowers said. "I drove it many miles. It was my primary transportation in high school."

The roots of what Flowers would later call his "genetic opposition to violence" were laid down three years later. With four friends, he set out one day in his buddy's fancy new family car, a red 1959 Plymouth Fury, to nearby Tulos, 16 miles away. They were heading southeast on state highway 127 when they spotted another car coming toward them—fast. The Tulos cops were right behind it, in hot pursuit.

The suspect's car hit Flowers and his friends' car nearly head-on, at 100 miles per hour. Two of Flowers's friends and one of the suspects were killed.

"I was lucky," Flowers said. "I was sitting in the backseat, on the side opposite the major thrust of the collision. I was conscious soon after the impact and remember amazing things about that experience. One of my friends died as he lay on top of me. I could not see or move much. I thought my left leg was gone but found my left foot over my right shoulder and could tell that it was still connected, which seemed strangely comforting." Flowers was last in the triage line and was taken to the nearest hospital in a station wagon. All the available ambulances were being used. He took 400 stitches in his face.

Today, Jena, Louisiana, has a recreation center named after Woodie Flowers's friends. And Flowers has a fierce, vocal loathing of any spectacle that involves crashing pieces of machinery into each other with deliberate force.

After he recovered from the accident, Flowers planned to become a professional welder in the Texas oil fields. But college—the Louisiana Polytechnic University—beckoned instead. He graduated in 1966, then moved to MIT to get his master's and his Ph.D. under Bob Mann. While Mann was trying to drag design education back into the metal shop, Flowers, inspired by the Boston Arm, was finishing his doctoral thesis on a mechanical knee that allowed patients to manipulate the prosthesis like a real joint. "The amputees were essentially wearing robots," Flowers said. "That was fairly powerful. We had to build quadruple redundant safety systems to make sure the system was absolutely failsafe."

He earned his Ph.D. from MIT in 1971, joined the engineering faculty that year, and went to work as an assistant on Herb Richardson's teaching staff. Flowers loved Richardson's design competition, and with office mate David Margolis he spent long nights thinking over the annual challenge and building machines of his own to test the course.

Flowers took over the class as head professor in 1974, and over the next decade, with a combination of his southern charm and enthusiasm, he turned the 2.70 competition into one of the most formative and exciting experiences for students at MIT.

His first step was to ban the media from covering the contest. During one early year, when the machines raced each other powered by the spring from a single mousetrap, a local TV station had sent a camera crew to film a segment on the evening news. Flowers hated the result. "It was five little clips of nerds at play. It was just a very bad piece of information transfer," he said.

Throughout the seventies, he worked to make the challenges more complex and exciting. Instead of springs and mousetraps, he put powerful motors in the students' kits, which were donated by companies like government contractor TRW—companies that Flowers personally solicited over the telephone. And instead of the students' machines competing by themselves against the clock, he pitted them against each other on parallel tracks, and then, even better, designed challenges so that two machines entered the field at the same time and directly competed.

Flowers changed the contest challenge every year. That way, he said, no one, even him, would know the best solution. "We tried to make a deal with students," Flowers said. "We're going to pose a good problem and give them a good kit. We knew they were going to really work hard on it. We're not going to embarrass anyone, and whether they win or not is not going to be part of grade. The competition will be a celebration of learning that has occurred up to that point."

The students responded. By the early eighties, the 2.70 competition was the most popular class on campus. Each year, hundreds of students would sign up, many from outside the mechanical-engineering department eager to compete in MIT's vaunted intellectual Super Bowl. The night before the class was to begin, Flowers and his staff would have the uncomfortable task of whittling the size down to a manageable number.

The competition also became a campuswide festival. Flowers moved it out of the mechanical-engineering classrooms to 26-100, a physics classroom and the largest lecture hall on campus. The MIT chorus performed at the event each year, and fraternities would show up to cheer for their members, painting their faces and celebrating their brothers' achievements. The joke at MIT, everyone said, was that even though the university had a football team, 2.70 was its true homecoming game. Flowers himself was full of energy at these competitions, and fashionable as always, would preside over the event wearing costume hats, suspenders, and vests adorned with decorations.

This was the showman the rest of the country would meet. In 1981, Flowers got a phone call from Graham Chedd and John Angier, producers of the PBS television show *Discover the World of Science.* The cofounders of Chedd-Angier, a Boston-based production company, wanted to do a story on this crazy engineering contest all the students at MIT were talking about. Flowers declined the invitation—he still remembered the insulting news clip from a decade ago—but the producers visited Flowers's office and promised that their story would stress the pedagogic philosophy behind the design competition, which sounded much better.

So, camera crews followed several teams of students over the course of six weeks in 1981 as they prepared for the event. The challenge that year was as follows: Two machines began the match on opposite ends of a 10-foot board, each clutching a peg. At the start of the match, students activated their machine by pressing a button. The goal was to design a contraption that would get the peg into a single hole at the center of the field fastest. Every competitor had a different strategy, from arms that quickly unfurled toward the peg, to contraptions that slid past the hole and tried to block their opponent. *Discover the World of Science* host Peter Graves (aka Captain Clarence Oveur in *Airplane!*) introduced the segment as "one of the most intense one-on-one intellectual competitions on the calendar."

The PBS cameras returned to MIT in 1984. By then, Flowers had enhanced the competition even further: Students directed their wheeled contraptions around the field using joysticks, which were connected to the machines by wire tethers that hung from the ceiling. The objective that year was to build a machine that could herd a field of Ping-Pong balls into a trough.

Once again, the students responded in unique ways. Some built machines that flung balls into their troughs with spinning blades, like miniature "slab kickers." Others built rakelike machines that pushed the balls, while still others built dump trucks, to load up on balls and haul them across the field. Many builders decorated their machines; one was dressed as a papier-mâché chicken.

Again the TV show was broadcast on PBS stations, and in thousands of schools nationwide that used the program as part of their science curriculum. An estimated three million people—from science students to gearheads—learned about the MIT competition, and its smiling, passionate ringleader, Woodie Flowers. "It wasn't just a sporting event. You always learned something about what the students were up against," said producer Joe Blatt, who worked for Chedd-Angier at the time. The PBS crew has returned nearly every year since 1984 to record the MIT event.

In the 1987 broadcast, a beaming Flowers addressed the cameras directly. "[Graduates] come back and say, Woodie, life's just a big 2.70 contest. It's the same stuff. You never have enough information. You never have enough time. The kit of materials may be what you got in the warehouse, but it's always constrained. There are always other people doing competing things and you must have a strategy.

"We've created a microcosm of the real engineering experience."

Woodie Flowers ran the MIT 2.70 contest for 13 years before handing the reins over to colleague Harry West in 1987. In 1989, working with professor Masashi Shimizu at the Tokyo Institute of Technology, West brought the competition overseas. A group of 12 students from that year's 2.70 class traveled to Tokyo. With another dozen Japanese students, they received a new kit of parts and a challenge similar to the one they faced that year at MIT. But instead of country-versus-country, Olympics-style, each American student joined with his or her Japanese counterpart. They would have to surmount cultural and language barriers, as well as engineering ones.

Both PBS and the Japanese public television network NHK broadcast the event that year. An NHK producer dubbed the contest "RoboCon." It simply sounded good, even though these kit-built, Ping-Pong-herding machines certainly weren't robots under academic definition, which decreed that robots needed onboard decision-making capabilities. From that day forward, students around the world who competed in task-based design competitions would call their creations "robots."

As for Woodie Flowers, fueled by the enthusiasm and attention around 2.70, his career took off. In 1990, a switch in underwriters forced Chedd-Angier to change the name *Discover the World of Science* to *Scientific American Frontiers,* and to revamp the show. They asked Flowers to be the host of the new program, and from 1990 to 1993 the professor, inventor, and contest impresario was also a TV star. "We really enjoyed his southern, down-home, honest way of speaking, and his surprisingly coy way of making some really profound points," said producer Joe Blatt.

In accepting the gig, Flowers told the MIT student newspaper, "We have a TV show called *L.A. Law,* but not *L.A. Engineer.* Teenagers can name any number of sports figures as heroes or role models, but they would be stumped if you asked them to name any scientists."

It was one of the first salvos in his campaign to raise the profile of science education. "I want to do things on the show that will allow viewers to make legitimate connections between excitement and careers in technology and science," he said. In his first show, Flowers proved he was serious. He jumped out of an airplane from 12,000 feet, falling a mile before deploying his parachute, and illustrating the principles of acceleration and weightlessness.

Woodie Flowers was almost 50 years old, but this was only the beginning of his crusade. Several months before, Flowers had met a flamboyant inventor from New Hampshire named Dean Kamen. Flowers had invited Kamen to speak to his product-design class, but Kamen was even more of a showman than Flowers, and wanted to land his helicopter right on MIT's campus. Flowers recalled the incident with a chuckle. "There was no way in hell MIT was going to let that happen."

Dean Kamen liked to make great entrances. He solved the world's intractable medical problems, crusaded for science education, warred against a culture that he considered inordinately focused on knuckleheaded athletes—and he made great entrances, usually inside or atop an invention of his own design. On his way to meet Woodie Flowers for the first time and to speak to his class in February 1990, Kamen figured he would put down his customized Enstrom 480 helicopter on the MIT campus and avoid Boston's infamous rush hour. After all, avoiding traffic was partly why he moved to bucolic New Hampshire—time was his most precious commodity. He asked Flowers to check if there was a place to land.

Flowers called him back. "The city of Cambridge won't allow it," the professor said.

"Woodie," Kamen replied, "isn't it surprising to you that MIT advances itself as this great, unique leader in the world of technology, and they have no place to put the coolest piece of flying technology ever invented?"

Kamen finally settled for parking his aircraft atop a local hospital and braving the mean streets of Cambridge in a taxi. But the episode was classic Kamen—always disgruntled with the way things are, never hesitating to vociferously challenge it.

His biggest challenge yet to the status quo would spring directly from that meeting with Woodie Flowers. It would take the form of the FIRST high-school robotics competition, an amped-up version of Flowers's contest—2.70 squared.

ꝺ

Dean Kamen was born in 1946, in Rockville Centre, Long Island, about 25 miles from Manhattan. His father, Jack, was a comic book artist who drew for *Mad* magazine (and according to legend, based the iconic Alfred E. Neuman on the irrepressible young Kamen himself). His mother, Evelyn, was a high-school teacher who would later keep the books on Kamen's many entrepreneurial efforts.

His father liked to refer to his middle son as the "human irritant," and said that, when Dean was four, the family learned they couldn't argue with him, since nothing they said was going to change his mind.

Kamen did poorly in school; instead of doing homework, he preferred to read the original works of scientists like Archimedes, Galileo, Newton, and Einstein. In his family's basement, he tinkered, building audiovisual equipment for sound and light shows and earning 60 grand a year before graduation. One summer, he worked on installations for New York's Hayden Planetarium; another year, he sent his parents on a cruise, then expanded the foundation of the house so he could add a machine shop in the basement. His parents came home when the project was only halfway completed, and heavy earth-moving equipment still sat in the backyard. His father looked at the mess behind the house, then at his son, then at his wife. "What's for dinner?" he asked with a shrug.

After stumbling through high school, Kamen enrolled at Worcester Polytechnic Institute. But he found himself drawn home every weekend to work on his newest project. His older brother, Bart, was studying to be a pediatric oncologist and noticed nurses were spending too much time checking patients' IVs to ensure they were administering proper dosages of medicine. Bart asked his younger brother, the mechanical genius, to invent a device that would automatically dispense small doses throughout the day.

Instead of attending college classes, Kamen went to work on the world's first portable drug infusion pump. He called his finished product Auto Syringe and recruited his mother, younger brother Mitch, and Mitch's friends to assemble the devices. By 1975, at the age of 24, Kamen had dropped out of college and moved his small company into a Long Island industrial building. A few years later, he invented a new device, the world's first portable insulin pump. Diabetics, who had been tethered to refrigerator-size machines at the hospital, could wear it on their belt and roam free.

By 1981, Kamen had sold Auto Syringe to international manufacturer Baxter and taken his riches to Manchester, New Hampshire, to escape the congestion and taxes of New York state. He moved his parents along with him. With this new windfall, he also fulfilled a childhood dream—to hover, to defy gravity—and bought his very first helicopter.

ᗡ

There are many other pieces of the Kamen story. He actually bought the company that made his helicopter, Enstrom, reworked its designs, then sold it again. He bought North Dumpling, a three-acre island in Long Island Sound, and conducted a mock secession from the United States over a squabble with the Department of the Interior, which had objected to his construction of a windmill. Kamen actually designed his own stamps and currency for the island and carried one of the bills around in his wallet.

In 1982, he started a new company, DEKA (short for his name), and set its offices inside a series of renovated nineteenth-century red brick textile mills along the Merrimack River in Manchester. While renovating those 200-year-old buildings, he ended up inventing a new industrial heating and cooling system and profitably spinning the technology off in a new company. In the mid-nineties, Kamen also invented a portable kidney dialysis unit, untethering kidney patients from 200-pound hospital machines.

He was a brilliant inventor and a shrewd businessman who knew how and when to spin off inventions to larger companies and return to doing what he did best: innovating. By all accounts, he was also an intense, driven boss, capable of bulldozing his staff. "Dean is so intense and so aggressive that you always have to worry whether he'll get frustrated at not moving fast enough," Bob Johnson, a former scientist at Johnson & Johnson, told *Wired* in a September 2000 profile. "Sometimes his intensity is almost frightening."

In his personal life, Kamen was frequently compared to Bruce Wayne, the eccentric millionaire bachelor living on a hill. He never married—girlfriends quickly tired of his workaholism. He designed and built a huge, hexagonal house outside Manchester and named it Westwind. A wind turbine provided extra power, a pulley system delivered bottles of wine to the bedroom, and his collection of antique machines included a jukebox, a slot machine, and a 16-ton steam engine once owned by Henry Ford. In the basement, Kamen kept everything a blue-blooded, grease-stained gearhead could want: a fully-stocked machine shop and foundry and a computer room with all the latest gadgets.

Outside the estate, Kamen had a full-size baseball field with night lighting, a basketball court, a tennis court, and two garages. In one, he kept a Porsche 928 and a black Hummer. In the other, he maintained two Enstrom helicopters—with modifications of his own design—and flew them five minutes to work, or to a nearby airfield, where he parked his Citation jet. The residents of Manchester called him "King Dean."

In the mid-1980s, Kamen decided he wanted to give something back to society. He was successful beyond his wildest dreams, with more than 100 patents and no shortage of new ideas. But he wasn't the kind of guy who could simply cut a check for someone else's charity. "I'm not good at it," he said, "and it doesn't give me the personal satisfaction." So, with the idea of making a small dent in the massive problem that was the declining science literacy of America's children, he built the SEE Science Center—an interactive engineering museum for kids—and put it on the first floor of one of his mill buildings. "We had a great time setting it up," Kamen recalled. "I would buy the pizza and beer, and my engineers and I would get together to figure out the kinds of things you put in a science center." The finished museum featured exhibits like the "moonwalk," a lunar landscape where kids wore a harness that reduced their weight by more than 80 percent, and a Bernoulli Blower—a ball that floated over a cone atop a stream of air.

Kamen felt proud of SEE, but there were still a few open spaces left in the science center, and one Saturday morning in 1989, he walked through it on his way to his office next door. There were kids and their parents there, playing with the exhibits. Why not ask them? Kamen walked over and inquired of his young customers what else they would like to see at SEE; they couldn't answer him. Then he asked if there were any particular scientists or inventions they wanted included. The kids didn't seem to know any scientists, except for Albert Einstein. Then Kamen asked the parents, and they couldn't answer either. Meanwhile, the inventor noticed, many of the kids wore Boston Bruins shirts and Red Sox caps—typical garb in sports-crazy New England.

Kamen walked back to his office, suddenly feeling bad about his museum. "Why was I so excited to build this place?" he said. "I'm sure it's doing good for the community, but it's probably doing more good for me. It's a place where the yuppie parents come and spend quality time with their kids. It's not going to spontaneously attract all the kids who need to understand that life isn't about dribbling."

Kamen concluded that what educators really faced in their quest to raise the popularity of science and technology among kids was the opposition of American culture itself. From the moment kids turned on a television, they were taught to venerate athletes. They watched Michael Jordan on the basketball court, making millions of dollars, while billboards screamed from every perch: "Be Like Mike," or "Life is Short, Play Hard." Kamen said, "Who is going to tell kids, life is short, work hard, or it's going to get a lot less shorter and a lot less fun?"

That's when he realized that his science center was great, but it wasn't going to cut it as his contribution to society. If he was going to demonstrate that science and technology could be as exciting as football or basketball—and a whole lot more likely to constitute a successful career—he would have to use the culture's own weapons against it. He would have to create a televised sporting competition, where the scientists and engineers were the athletes, having fun, playing hard, and reaping the benefits of success. It would have to be, he decided, an Olympic Committee of Smarts.

The idea started to take shape later that year, when Kamen stayed overnight at the Gainesville, Florida, cabin of longtime friend and mentor William Murphy. Always searching for good engineers for his various high-tech companies, Murphy had tried to recruit Kamen when he was in his twenties. He soon realized that the young inventor marched to his own beat, but he was so impressed with Kamen's intensity and talent that they stayed in touch.

That evening in Gainesville, Kamen and Murphy commiserated about how difficult it was to hire competent engineers. Over the course of the conversation, Kamen asked to see the Nobel Prize won by Murphy's father, William Parry Murphy, who had achieved breakthroughs in the treatment of pernicious anemia in the 1920s. Murphy told Kamen that the award was in a closet. He didn't know what to do with it.

Kamen couldn't believe it: "This guy has a Nobel Prize sitting in the closet, and the jockstrap of some athlete is hanging in a plastic case somewhere?" he marveled. Lying in his bed that night in Murphy's cabin, he decided the first component of his Olympic Committee of Smarts would be an Engineering Hall of Fame. Soon afterward, he gave his new philanthropic effort a formal name: FIRST, For Inspiration and Recognition of Science and Technology. "The media have created megaheroes of rock stars and sports players," Kamen said. "We were trying to figure out how to make the next hero a scientist or a technologist."

All he needed now was his own spark of inspiration, to give direction to his vague notions of creating an engineering-based sport. Then, early the next year, 1990, he flew his helicopter to a Cambridge hospital and shuttled over to the MIT campus to speak to Woodie Flowers's design class.

It was a meeting of the minds: Flowers himself had recently gagged on the news that Boston was naming its new tunnel underneath the harbor after a hot-hitting left fielder named Ted Williams. "I was stunned, I thought I must not have heard what I thought I heard," Flowers said. Kamen wholeheartedly agreed. In the Greater Scheme of Things, did ballplayers who hit over .400 in one season really mean more to society than the genius gearheads who revitalized the lives of diabetics, or gave amputees control over their artificial knees? "The culture gets what it celebrates," they both said, so it was no wonder the country ranked fourteenth in science education.

After the meeting, Flowers e-mailed Kamen his thanks for the speech to his students and offered to talk to MIT's president about Kamen's plans for an engineering hall of fame. Flowers also mailed Kamen a copy of the *Discover the World of Science* presentation of the 1987 2.70 competition.

"Bingo," Kamen thought while watching the video. This would be his sport. He would take the format of the 2.70 competition—and square it. Instead of college students working alone to build a robot, real engineers from real companies would participate, working hand in hand with teams of high-school students. Instead of a small kit of rubber bands and tongue depressors weighing 10 pounds, Kamen would solicit robust mechanical components to put into kits of 50 to 100 pounds. The robots wouldn't climb over tables or crawl up strings, they would be tall, mobile, kick-ass machines on wheels.

Kamen knew he couldn't do it alone. "Woodie had spent his career making those games work," Kamen said. He invited Flowers to his home in New Hampshire. He didn't realize the professor would melt when told the idea, and started aggressively laying out his vision. "Woodie, the fact that you can take students that are already at MIT and convince them that doing engineering is fun is like carrying coals to Newcastle," he told the professor. "Kids like to play basketball because they see adults playing basketball. I want to get the pros involved in something that can be presented as a sport, and get kids to watch."

Flowers jumped right aboard. They would make a peculiar couple: the tall, lanky, pony-tailed professor from the South, and the short, energetic inventor from Long Island with the helmet of black hair and a messianic streak. The FIRST robotics competition was on its way.

The next step was to find the professional athletes for his new engineering sport. They were hiding, Kamen knew, inside corporate America's biggest companies. In 1991, Kamen addressed a high-technology consortium called the Council on Competitiveness.

Speaking in front of luminaries like Paul Allaire, chairman of Xerox, George Fisher, chairman of Motorola, and Frank Shrontz, chairman of Boeing, Kamen told the tech titans to stop blaming others for the sad state of American science education. After all, he said, your companies are the ones sponsoring sporting events like the Olympics and pro basketball. "These are the distractions that just totally obscure to kids what's important in their lives," Kamen said. "Why can't you guys get together and sponsor something that teaches kids the things you want them to know, and gives them the aspirations that you'd like them to have?"

The inventor later recalled, "They agreed on the spot. I got about 20 of those companies." Each ponied up $3,000 for the first year of FIRST, and assigned their engineers to take a high school under their wing.

So while Marc Thorpe was just beginning to sketch out his ideas for robotic combat in California, the inaugural task-based robot competition for high-schoolers was held in February 1992, in the Memorial High School gymnasium in Manchester. Twenty-eight teams participated, each of them composed of high-school kids working together with professional engineers from firms like Alcoa, AT&T, Boeing, and Xerox.

The students drove the robots over a 16-by-16-foot arena, covered with whole corn kernels, which provided an extra traction challenge. The 24-pound robots had to collect various colored tennis balls placed in a circle at the center of the field, with each color representing a different point total, and herd them into their corner. Four robots competed against each other in two-minute rounds.

At the event, a bearded, vigorous Kamen was practically bouncing off the walls, preaching to anyone with a pen or a TV camera. "We took [science education] out of the dry, noncompetitive, rote exercise of doing boring things, and put it into an arena kids can understand: competition, with winners and losers," he said.

He also initiated a new practice—pestering politicians to support his cause. George Bush Senior was campaigning for reelection in New Hampshire at the time, and the president swung by the high-school gym to watch the robot showdown. According to a profile of Kamen that appeared in *Smithsonian* magazine later that year, Kamen relentlessly dogged Bush, pleading for his support. "Mr. President, you're always inviting the winners of football games and basketball games to the White House. Why not encourage young scientists by inviting the winners of FIRST?" Bush politely agreed, then tried to walk away. Kamen followed, tugging at his jacket and urging him to walk to the microphone and announce it right away. Bush did.

In subsequent years, Kamen would harangue countless other politicians and corporate titans with the urgency of FIRST's mission. "I never miss the opportunity to bludgeon somebody into supporting us," Kamen liked to say. President Bill Clinton, himself the owner of a legendary reserve of stamina, admitted to meeting his match in the inventor. "His energy is the single most inexhaustible thing I think I've seen in the United States of America," Clinton said. "If you do not want to hear about what he does, do not ask, or stand within a four-mile radius." Not surprisingly, FIRST winners were invited to the White House throughout the early nineties.

In 1994, 43 teams from 15 states gathered in the New Hampshire gym, with 65-pound robots that lifted soccer balls and placed them into goals five feet high. Two thousand people turned out to watch the third annual competition. "In 10 years, this will overshadow the Super Bowl," Kamen predicted. "We will graduate to where FIRST gets national media attention, people follow the season of events, and it develops all the trappings of other professional, collegiate, and high-school sports.

"Our goal is to change the culture of America," he said, and he couldn't have been more explicit: Kamen wanted to get FIRST on television. "I don't want it to be a festival that no one but the nerds knows about," he said. "The goal is to get it on TV so that thousands of kids can see it."

By 1994, he was already achieving some success. Kamen was getting ESPN to record and broadcast highlights of the event, although at odd hours of the day and to small audiences. Still, it was a start, and Kamen felt the exposure would help introduce the competition to the public and pave the way for a real TV exposure later.

There was another goal for FIRST, and here Woodie Flowers was a strong influence on Kamen. The competition, the professor urged, should inspire certain values and ideals that the students would need in the real world. Flowers talked about these precepts each year, particularly when the leaders of each team met at Kamen's estate in New Hampshire for the annual kickoff. There they received their kits and began six weeks of preparation for the tournament. Flowers emphasized that FIRST was about nonviolence, an "everybody wins" ethos, and cooperation between competitors. "We're not necessarily genetically programmed to need violence and losers in order to find victors," Flowers often said.

One year during a speech, Flowers displayed a slide labeled "gracious professionalism." It was about everybody winning and helping each other. Kamen loved that, and the mantra soon pervaded all of FIRST's literature.

Meanwhile, the impressive grass-roots growth of FIRST continued. In 1995, Kamen introduced two regional competitions and moved the national event from New Hampshire to an auditorium inside Epcot Center at Disney World. By 1997, there were eight regionals around the country, and the nationals grew so large they were moved to the parking lot outside Epcot. From 44 teams in 1994, the competition grew to 93 teams in 1996, and more than 100 in 1997. The robots now weighed over 100 pounds.

FIRST events had all the trappings of a sporting match: The teams of students and engineers dyed their hair, brought mascots, and decorated their robots. They wore team T-shirts and treated the game like a true athletic competition. The sponsors continued to line up: NASA, GM, automotive parts maker Delphi, Disney, and many others poured thousands of dollars into FIRST teams, and their engineers spent countless hours over six weeks helping students build their robots for the competition

During the competition Woodie Flowers was the consummate entertainer. He MCed the events wearing a Dr. Seuss hat one year, a cowboy hat another, a Viking hat another. He put on every pin, button, and T-shirt the students gave him, and if he saw a group of neglected or stressed-out students, he gently engaged and encouraged them. "This has a chance of really mattering," Flowers said. "We might be able to change millions of kids' attitudes about education." At the tournaments, the students treated Flowers and Kamen like heroes, asking them to sign autographs and to pose for pictures.

But the best sign of all was that there was no shortage of FIRST graduates who went into college engineering programs and later testified that the robot competition had changed their lives. Kamen toted these stories around with him and always made sure to mention them in his speeches. He was a proud papa.

¤

The first real indication that Kamen's two goals might be incompatible came in 1999.

That year, he lured Nickelodeon, the children's TV network, and a division of media conglomerate Viacom, to film the FIRST competition. This was more than 15 months before *Battlebots* would debut on Comedy Central, about a year before Viacom's MTV division would film a *Robot Wars* pilot in London.

Two Viacom execs, Kevin Kay and Albie Hecht, came out to the FIRST nationals at Disney World with an army of cameras. Finally, here was the potential for the realization of the key component of Kamen's dream—to get FIRST on TV. Camera crews swarmed the pits, where 15,000 students and mentors toiled over 270 robots.

The challenge that year was distinctive, and it represented a new trend for Kamen and Flowers as they attempted to drive home the values of nonviolence and gracious professionalism. Instead of three or four teams working against each other as in past years, they instituted a system of alliances. In every match, each team would collaborate with another team, working in tandem against another alliance of two. The game, called "Double Trouble," involved the robots' picking up rubber balls and dropping them into bins. The scoring system was complex and unintuitive: Robots scored one point for the yellow balls, five points for the scarcer black ones. They also got points for ascending a ramp in the center of the field, hanging from a chin-up bar, and helping their partners hang from the bar.

Kamen and Flowers added one more flourish to underscore the point: The winning alliance, they decreed, would get a final score three times that of the losing alliance. This created a strong incentive for powerful teams to avoid crushing their opponent.

1

Before seizing upon the idea of robotic combat, Marc Thorpe studied dolphin behavior and worked on special effects at George Lucas's Industrial Light & Magic.

2

3

Mark Pauline, the defiant, iconoclastic founder of Survival Research Labs, momentarily checks his disdain for authority as he squares his plans for a performance with the SFPD.

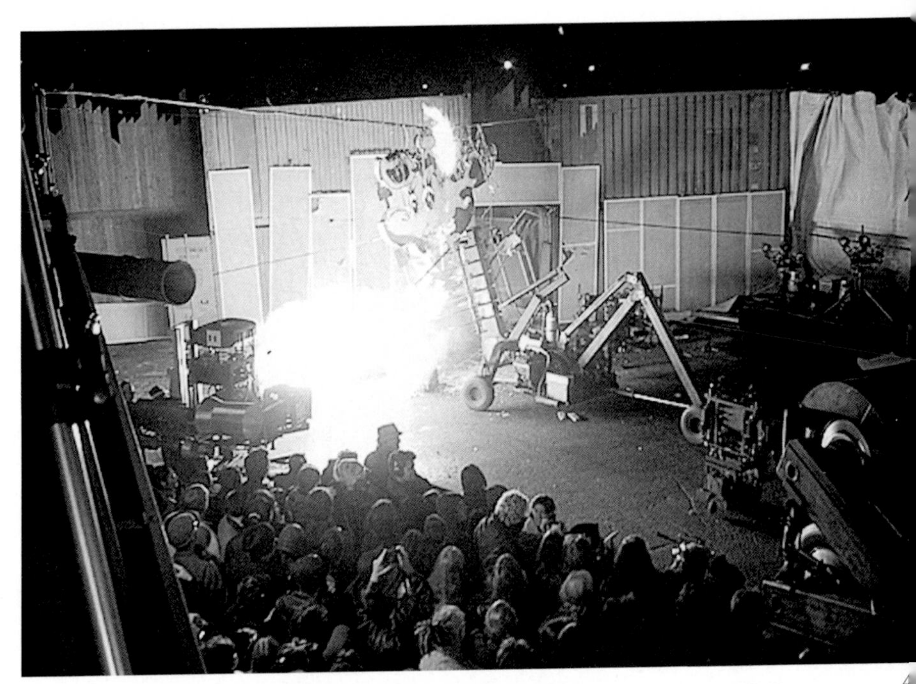

Mayhem 'R Us: An awed crowd looks on as a fire-belching SRL monstrosity annihilates its mark, Geoffrey the Giraffe.

4

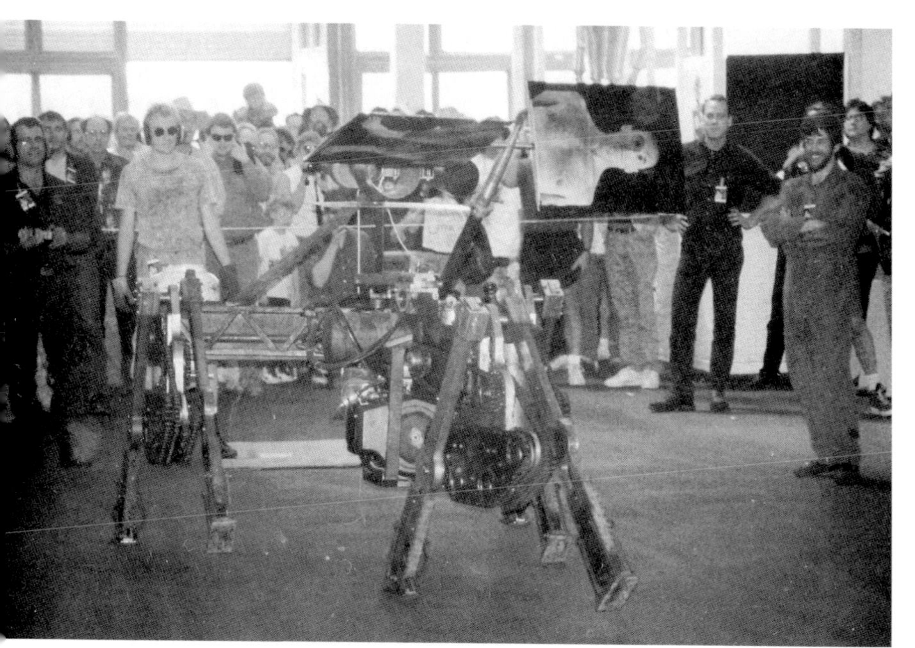

5

Although he agreed to demonstrate an SRL robot at the event, Mark Pauline alienated spectators at the inaugural Robot Wars competition.

6

"Supreme Commander" Charles Tilford plays to the media crush at Fort Mason.

The arena used for the first Robot Wars hardly sported the sort of insidious booby traps that today's intrepid robots must brave.

7

8

Will Wright, creator of the Sim family of video games, tends to JulieBot, while another competitor looks on.

AndyRoid and Spiny Norman, two, uh, *unique* fan favorites, duke it out at Robot Wars '94.

9

Posters for Robot Wars '94, '95, and '97.

10

11

12

13

Trey Roski's enthusiasm for the new sport has survived battles in both the robotic and the business arenas.

14

Sure, it may not look like much, but Blendo's wicked kinetic energy made it the most feared bot in the biz—among competitors and organizers.

15

According to *The San Francisco Chronicle,* Thor hopped around "like a piece of horny farm equipment" when clobbering opponents with its hammer.

16

Carlo Bertocchini poses with his uber-champion robot, Biohazard.

17

By the third annual Robot Wars competition, more and more people were being drawn to the new sport of robotic combat.

18

Though Team La Machine fell to Biohazard in the heavyweight championship match, Trey Roski and Greg Munson prevailed in the next day's rumble.

19

There are few places as frenzied as the pits at a robotic-sporting event, but the competitors almost always take time out from tinkering with their own warriors to help out with someone else's bot.

20

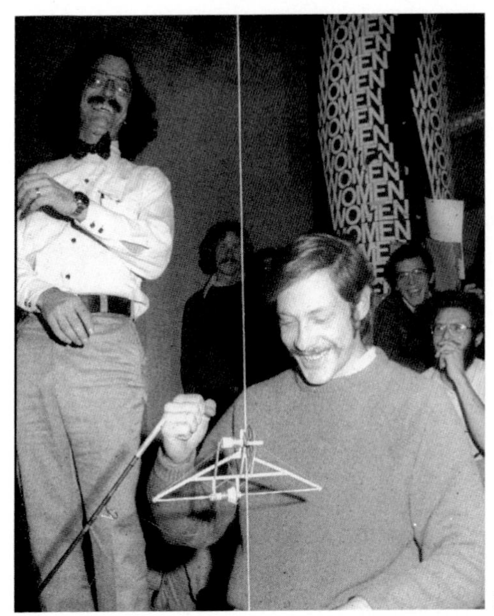

21

Under the direction of flamboyant MIT professor Woodie Flowers (left) the annual 2.70 engineering design contest graduated from campus phenomenon to televised spectacle in the interest of science education.

22

Inventor, entrepreneur, and self-styled visionary Dean Kamen demonstrates his Segway Human Transporter.

23

A study in contrasts: As a group of high school students work together to navigate their robot through the completion of their decidedly nonviolent task at the annual FIRST competition (above), the spinner Ziggo (background below) surveys the utterly mangled remains of its Battlebots opponent.

24

Sir Killalot, the most popular and fearsome of the house robots, terrorizes less impressive bots on *Robot Wars UK.*

Special-effects guru Mark Setrakian displays Mechadon, his latest and most ambitious robot, to an amazed crowd. More than any other builder, Setrakian has advanced the technology and artistry of robot combat.

26

27

Trey Roski dramatically enters the Battlebox on his new robot, Ginsu.

28

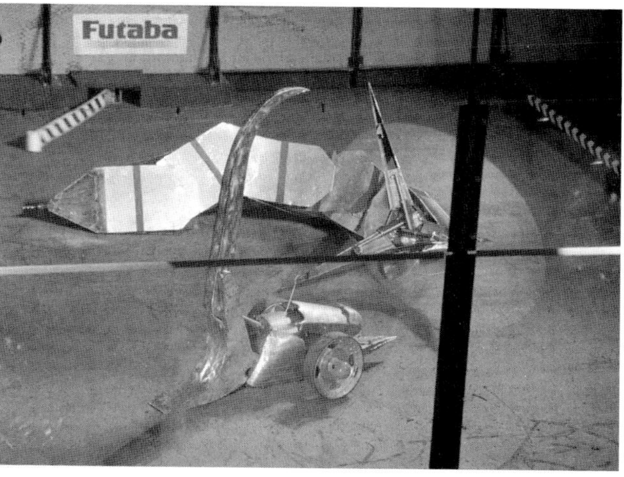

Setrakian's Snake faces off against Scorpion; although Snake was a feat of creative engineering, it was vulnerable to attack from more conventional bots.

2

Son of Whyachi and Nightmare clash in the center of the arena. Team Whyachi was reviled by some gearheads for their flashy, NASCAR-inspired uniforms and obviously deep pockets.

30

In February 2002, competitors gathered to celebrate the creative resolution of Thorpe's legal troubles by his attorney, David Chandler (left). But the sting remains for Thorpe.

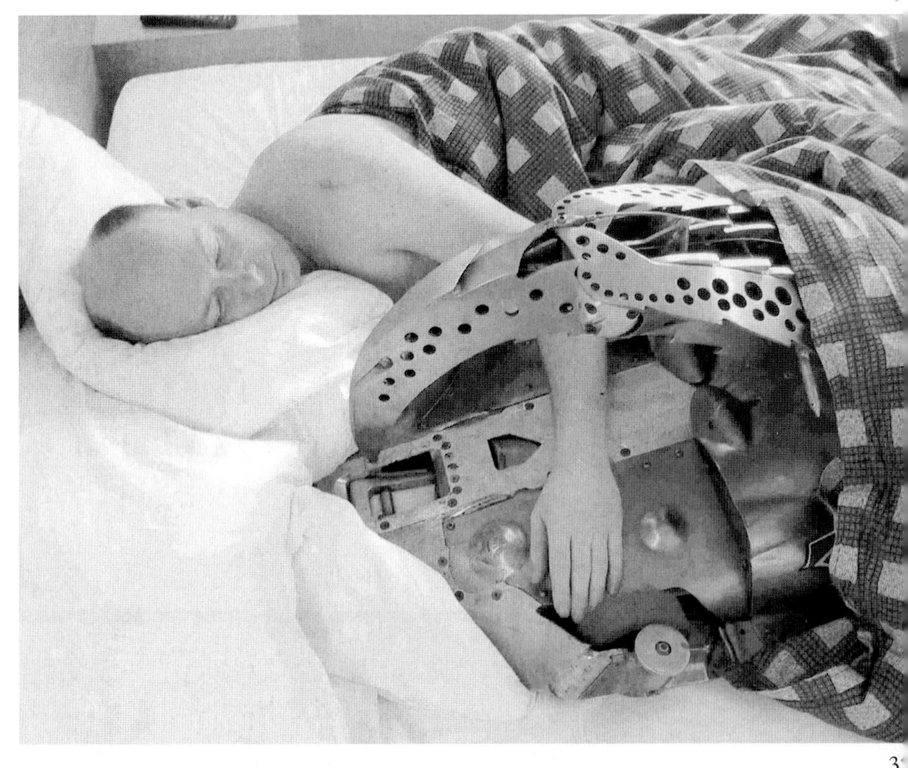

Have you hugged your robot today?

Kamen called the new format "coompetition"—and argued that it mirrored the way teams of engineers worked in real life.

After the event, Kevin Kay and Albie Hecht took their footage of the FIRST competition back to Nickelodeon. They edited the tape into a pilot—and it never ran. Kevin Kay said that the FIRST game was simply not made for television. "Everything they're doing from a values standpoint is worthwhile," he said. "But we decided that the things they have their robots doing aren't exciting enough. Ultimately, it's not great TV."

Ironically, the passion of the students made the Viacom execs more confident than ever about the possibilities of televised robot competition. After they discarded their FIRST pilot, Kay and Hecht began exploring other ideas for a robot television show. They met with Trey Roski, who was trying to find a home for *Battlebots* at the time, and when Roski ultimately went with Comedy Central, the partners settled on another, more accessible property within the Viacom family. By then, MTV had passed on its pilot of Mentorn's *Robot Wars.* Kay and Hecht snapped up the option and went to work finding another home among Viacom's networks for an American version of the British game show, featuring the fearsome and indestructible house robots.

Kamen didn't know any of this, nor did he know about the long trajectory of Marc Thorpe's Robot Wars through the U.S. courts. In fact, he was riding high: President Bill Clinton had just presented him with the 2000 National Medal of Technology for the iBot, a six-wheeled wheelchair with a series of gyroscopes that enabled it to climb and descend stairs and lift passengers to eye level. And he was a year away from springing a new invention on the world, a top-secret, dual-wheeled transporter known around DEKA only by the code-name "Ginger."

But four months after the 2000 FIRST nationals, Trey Roski's brand of robotic combat became the first robot competition to debut as a television show in the United States. The format could not have been simpler or more intuitive: robot in blue corner; robot in red corner; fight.

Robotic competition had finally debuted as a regular American sporting event, venerating the gearheads and engineers on a national network—but it wasn't FIRST.

ꟼ

With millions of other Americans, Dean Kamen sat down to watch the debut episode of *Battlebots.* He heard cohost Sean Salisbury tell the audience, "We are committed to restoring your faith in uncensored acts of extreme violence," and he watched as cohost Bil Dwyer pitched in, "Tonight, right before your very eyes, you are going to see real live robots fight to the death."

Then Kamen listened as Morgan Tilford calmly informed the audience: "You've got to make the other robot your bitch."

To Kamen, *Battlebots* was a horror. With only the Comedy Central presentation to judge from, he concluded that it celebrated all the wrong things—bad behavior, bad sportsmanship, and bad values. To Woodie Flowers, who also watched down in Boston, *Battlebots* was an abomination. Here was the ultimate manifestation of his distaste for dangerous machines in the hands of naïve designers. *The robots were deliberately crashed into each other.*

Then it got worse. In an unfortunate coincidence, the company that produced *Battlebots* in conjunction with Comedy Central was called First Television. Naturally, people began confusing *Battlebots* with FIRST. Kamen had to explain to friends at his parents' country club about the fundamental differences between the two competitions. While soliciting sponsorships from such companies as Ford, Kamen found himself telling execs that, no, they hadn't seen the FIRST competition on television last night.

The inventor had worked far too long and too hard to be coopted by this new, lowest-common-denominator competition, which had seemingly come out of nowhere. Later that year, while preparing the challenge for the 2001 FIRST competition, Kamen huddled with Flowers, and the partners devised their response. It would be a direct repudiation of the base morality exhibited in robotic combat. They would prove that you don't need violence in competition, that working together, as a team, is just as exciting.

As they did every year, they announced the 2001 challenge at Westwind. Kamen traveled among his guests sitting in the iBot.

"In this year's game," Kamen told the crowd, "I think I made the most diabolical change I've ever made. There is nobody to crash and bash or defeat. In this year's game, everybody succeeds together or everybody suffers together."

As it had the previous two years, the challenge involved four robots on the field at the same time. But the four teams wouldn't compete with each other in teams of two, they would all collaborate in one giant alliance of four robots trying to score the highest point total they could. The game involved balancing the robots on a seesaw at the middle of the arena and lifting soccer balls onto tall goals.

At the events that year, Kamen made sure that the world knew his agenda. "This year, we said we're going to put everybody on the same team, so they get out on that field and there is no mindless violence. There is no reason for any team to do anything but help everybody."

The subtext was plain: FIRST wasn't Battlebots. It wasn't about destroying, but strategizing and working out problems together.

In their speeches at the 2001 regional and national events, Kamen and Flowers didn't explicitly mention Battlebots, but their point was plain to everyone. Robotic combat belonged in the giant hamper labeled: "things wrong with America."

Most of the students hated the new, everybody-wins format. They took their strong feelings to the FIRST community bulletin boards on the Internet, and they worried about the future of their sport.

"Frankly, I feel such immense disappointment that I fear it will be hard for me to even enjoy any of the 'competitions,'" wrote one student. "I understand that FIRST wants to glorify the idea of cooperation, but I feel competition is necessary to keep public interest high and to ensure growth."

"I know that Dean and company believe that Battlebots is all about violence and celebrating what's wrong with society, but many people fail to notice the positive aspects of Battlebots," said another, who went on to note the sense of community and camaraderie in both pits. "The pits at a Battlebots competition are much like a FIRST competition. People are looking to help others."

"Where's gracious professionalism if we can't at least be open to other forms of robot competition?" asked another student.

Many others wondered if Kamen was losing touch with the real competitive desires of kids, and they lamented that Battlebots had jumped on TV first, even though FIRST and task-based robot competitions had been around far longer.

"We can't even think of gaining any of their audience because there's not much to watch . . . the games are hard to understand . . . not easy to score," wrote one poster.

Another active poster named Dan Haeg, a student at Carnegie Mellon, had participated in both events. "Ironically," Haeg wrote, "*Battlebots* will do what Dean and Woodie are trying to do with FIRST. It will turn engineers into stars. Kids will play with their *Battlebots* figures; they already do in the United Kingdom, where *Robot Wars* R/C cars and action figures are now available."

A minority on the bulletin boards joined Kamen and Flowers in their distaste for the *Battlebots* TV show. "My doubts began when I thought that it could have been me outside the Battlebox, being asked about my 'great shaft work' by the blond bimbo . . . er . . . reporter," wrote one student.

Still, to many of the FIRST participants active on the Internet, it seemed like another competition had eaten their lunch.

Into this collision of competing visions and values stepped Lenny Stucker. The TalentWorks chief and Battlebots VP was clueless about the schism in the competitive robotics world. Trey Roski and Stucker had discussed the possibility of spinning out a high-school version of their sport—to be called Battlebots IQ—and Stucker figured he would check out some of the other entities in the field, and perhaps explore the idea of a collaboration.

In March 2001, while Kamen and Flowers were making their most explicit statements yet about favoring gracious professionalism over violent combat, Stucker hopped a taxi up to Morningside Heights for the New York City FIRST regional. He didn't tell Roski or the rest of the Battlebots crew what he was doing. After hearing reports of Kamen's animosity from competitors who straddled both worlds, they undoubtedly would have warned him. Oblivious,

Stucker entered Columbia's Levien Gym, where 5,000 visitors were cheering on 40 local teams. Stucker asked around for the competition's organizers and was directed to Woodie Flowers, who was sitting on one side of the arena having lunch.

The TV sports veteran had dealt with some of the meanest, shadiest characters in the world of professional boxing, but he was about to meet his match in the pony-tailed professor. Stucker walked over and introduced himself as a Battlebots organizer.

"Hmm," said Flowers.

Stucker casually mentioned that the other Battlebots organizers were going to Boston that weekend, to talk about Battlebots IQ with Flowers's colleague, Alex Slocum, the mechanical-engineering professor who had taken over the 2.70 competition after Harry West.

Flowers leveled a stern gaze at Stucker. "I find it hard to believe that anyone at MIT would associate themselves with Battlebots," he said. "Now, if you'll excuse me, I'm having lunch right now."

Afterward, Stucker fulminated about his experience at the FIRST event. "These are stuffed-shirt, pompous people," Stucker raged. "Their sport is deadly to watch. Watching it is like watching paint dry. And they are so pompous that they wouldn't even listen to me."

Later, Roski tried to set up a meeting of peace with Dean Kamen. FIRST sent marketing director Laura London instead. Now Roski and Stucker both felt snubbed. They would develop their Battlebots IQ competition with a vengeance. Though it would take several years for it to get off the ground and wouldn't debut until early 2002, Stucker vowed, "Battlebots IQ is going to blow FIRST out of the water."

Mike Bastoni, a teacher at Boston's Plymouth high school who led teams of students to both Battlebots and FIRST, was more circumspect about what had transpired. "At a time that's inopportune for the development of robotic sports," he said, "we have this divide."

The tornado swirled for the remainder of spring 2001, and MIT professor Alex Slocum found himself at the vortex. The newest custodian of 2.70 had indeed met with Trey Roski, Greg Munson, and Nola Garcia—a longtime FIRST and Battlebots participant from Miami, who would spearhead Battlebots IQ—and he agreed to help them design the curriculum for their new high-school robotics competition. This would prove an unpopular decision on the fourth floor of MIT's mechanical-engineering department.

Slocum, a bearded, 43-year-old teacher and researcher, was unusually strong for an engineering professor: He had once hoisted two full-size college students into the air, one in each arm, and spun them in a congratulatory whirl after they had won the 2.70 contest.

Under Slocum's watch, 2.70 (the formal name of the course had actually changed to 2.007) had grown larger than ever, taking place each year in the university's Johnson Athletic Center. Chedd-Angier still filmed it and had expanded their annual robot competition episode to include other robot competitions, such as the RoboCup soccer tournament and the Trinity College Fire Fighting Robot contest. At MIT, the winning 2.70 students moved on to RoboCon in Japan, which by 2001 included 48 students from six countries.

Slocum possessed all the energy and showmanship of a young Woodie Flowers, and seemed willing to put his fingers into just as many pies. "I'm looking at the evolution of robot competitions in the spirit of design itself," Slocum said of his decision to work with the crew from Battlebots. "Let's see what we can do with it. It's stupid to throw stones at any of this stuff."

Slocum shook his head at what happened next. Woodie Flowers opposed any involvement in Battlebots and kicked up an interdepartmental fuss. The department head called Slocum into his office and demanded he explain his position. "I told him I was running an experiment in the true scientific method," Slocum said. "How can you condemn something if you don't have any data? I'm going to get data on Battlebots."

Slocum attended the next Battlebots competition in May 2001 and reported back that the camaraderie and spirit in the pits "was just like in other competitions. People share equipment and cheer on everyone else." Nevertheless, he decided to curtail his involvement with Battlebots IQ because of Flowers's opposition.

Although the two professors had the same goal—the elevation of the status of the engineer—the worlds of robotic sports remained far apart.

Slocum also had a difficult time understanding Dean Kamen's hostility toward gladiatorial forms of robot contests. After all, he said, wasn't business itself like robotic combat—a Darwinian struggle, where only the fittest survive? Kamen himself was a savvy businessman, whose job it was "to make money, make his employees money, and to put the other guys out of business," Slocum said. "People who live in glass houses shouldn't discharge automatic weapons."

Kamen angrily rejected that point of view; he was offended by it. "You think I'm going to sneak out and blow up my competitors' plants? Or that they are going to go out and destroy all the motor manufacturers so I can't build my products? I don't think so," the inventor argued. "The way the world of technology works is that I put out the best solution I can come up with, you put out your best solution, and the marketplace determines which is better. If you win, I'll be drinking beer and sending you champagne, and then I'll work on something else. You win by doing a better job at something, not by going out and killing somebody. FIRST is not Battlebots. It might seem like a subtle difference to some, but it's a big difference to me."

It all boiled down to a battle of metaphors: Was the world more like a FIRST competition, or a Battlebots competition?

Do people generally work together toward creative solutions, graciously competing and leaving the field inspired, empowered, and motivated?

Or, was the world darker than that, with winners and, more specifically, losers—like a penniless artist, who fought for his life in bankruptcy court, with nothing left to lose besides the friends who remembered all the great things he had done, way back when?

Chapter 9

Marc Thorpe vs. Robot Wars

Whatever the nature of the cosmic forces that brought Thorpe and Plotnicki together, not even the force of bankruptcy can tear them apart.

—Attorney William Weintraub, fee application before Santa Rosa bankruptcy court, May 2001

Marc Thorpe tuned in to *Battlebots* exactly twice.

He hated the sexual innuendo, the parody of basketball and football, and the strained effort to accentuate the eccentricity of the gearheads. "Robot Wars in the nineties was honest, direct, without hype," he said. "There was no heavy sales pitch. It was treated as homegrown and genuine."

There was also another, deeper reason he couldn't watch. The competition he dreamed up to secure his family's future had ruined his life.

After he settled his bankruptcy case in 1999, Thorpe's legal quagmire deepened. Steve Plotnicki felt that Thorpe had violated the terms of the settlement with his disgruntled messages on the Delphi forum, including the angry "Rodney Thorpe" missive. By attending the first Battlebots event in Long Beach, Plotnicki felt Thorpe was tacitly endorsing the competition.

The last thing Plotnicki was going to do was pay Thorpe the hundreds of thousands of dollars he owed him under the settlement agreement. Waging war against Thorpe in Santa Rosa could also be beneficial in another way: as a tactic in proving a conspiracy between Marc Thorpe and Trey Roski, which could then be used to restart the battle against the Roski family in New York federal court.

In Santa Rosa, Plotnicki subpoenaed all the e-mail that Thorpe had sent and received from the Robot Wars AOL account, and all the messages he had posted to the Delphi Web forum. He also subpoenaed all the Thorpe family's personal telephone records, credit-card bills, and all documents in Thorpe's possession that mentioned anyone on a list of 21 key figures in the robotic combat world, including Trey Roski and Bob Leppo. Plotnicki believed Thorpe had conspired with Roski to form Battlebots and circumvent Profile, and he was determined to root out evidence of the scheme.

In February 2000, Thorpe's attorney, William Weintraub, left the case when Trey Roski's $150,000 retainer ran out. Weintraub referred the Thorpes to David Chandler, a Santa Rosa–based graduate of the McGeorge School of Law in Sacramento, who was known in local bankruptcy circles as a legal pit bull, an attorney who preferred to wage war and win rather than settle a case.

Within the bankruptcy proceeding, Chandler initiated a lawsuit against Plotnicki, demanding that he pay Thorpe the cash and royalties delineated under the settlement agreement—a total he estimated at $650,000 and growing. In a counterclaim, Plotnicki asked to be relieved of the payment, arguing that Thorpe had broken the settlement with his Web postings and plotted with Roski to create Battlebots.

In court filings, Plotnicki revealed what he had found on his fishing expedition through the Thorpe family records: $23,124 in payments from venture capitalist Bob Leppo to Marc and Denise Thorpe. Leppo had previously loaned Thorpe money to hire an attorney and put on the 1997 Robot Wars. Still a believer in the promise of robotic combat, the gray-haired venture capitalist had also invested in Battlebots, and sat on its board.

Plotnicki discovered that during 1999, while Thorpe had a legal responsibility to promote Robot Wars and was posting his lukewarm messages to the Delphi forum, Leppo (now a Battlebots investor) was paying the artist thousands of dollars through a Marin County investment banker named Rand Selig. To Plotnicki, this was the "smoking gun" that proved wrongdoing.

Thorpe and Leppo explained that the funds were legitimate payment for a video, produced by Thorpe and his wife, called "So . . . What Are We Going to do with Your Hair?" Leppo had felt bad for his friend in the midst of his legal, financial, and medical woes. He asked him if there was any other project he could put money into, and he picked the beauty video—Denise Thorpe's idea—from a list of possible projects that Thorpe brought him. The concept was to try to match the success of a video called *Road Construction Ahead,* a 1991 documentary about roadwork that was a hit with construction-vehicle-obsessed young kids. The so-called "beauty video" followed the activity in a northern California salon and recorded tips from hairstylists and makeup artists working on young male and female customers. "Bob Leppo is both a friend and a businessman," Thorpe said. "It made good sense to him as a business investment."

Thorpe and his wife spent six months working on the documentary with Thorpe's old friend Greg Becker, the maker of the aborted video from the first Robot Wars. The video was released in 2001, and Leppo admitted, "We were never able to get sales going."

Steve Plotnicki rejected the whole explanation. He saw the payments as a tacit bribe, a hidden reward to Thorpe for breaking his settlement and refusing to promote Robot Wars. In court documents, Plotnicki argued Thorpe and his wife had no experience making these kinds of documentaries. (Despite their 1970s dolphin project, plus all of their work at ILM, where Dennie had also worked as a sound engineer.) He also pointed to phone bills that showed 200 calls between Thorpe and Trey Roski throughout the early months of 1999, before and after Battlebots was announced. Plotnicki felt that evidence justified his asking for relief from paying Thorpe any royalties or cash from the Robot Wars business.

It all came to a head in the Santa Rosa courtroom of Judge Alan Jaroslovsky. For Jaroslovsky, the crux of the issue was not beauty videos, hidden cash payments, or tacit endorsements of rival events. He wanted to know if Marc Thorpe had truly violated his promise to "use his reasonable best efforts to rehabilitate the reputation" of Robot Wars, and if he had, Jaroslovsky needed to decide whether this even made a difference in the outcome. In other words, was Thorpe responsible for killing Robot Wars in America, as Plotnicki claimed?

Or had Plotnicki hopelessly alienated members of the robot community himself, with his legal tactics.

To decide the issues, the judge heard two trials during the summer of 2000. The first, held in June, concerned Profile's contention that Thorpe acted in bad faith and violated the settlement. In this proceeding, Steve Plotnicki took the stand, and David Chandler cross-examined him for the entire morning.

Chandler grilled Plotnicki on why he hadn't paid Thorpe the $250,000 delineated under the settlement agreement. The lawyer wondered aloud if there was anything Thorpe could have done in terms of "rehabilitating" Plotnicki's reputation that would have truly satisfied him.

Later, Chandler asked Plotnicki why he was so sure Thorpe had conspired with Roski to form Battlebots. "His commercially unreasonable behavior for the past three years," Plotnicki answered. "Walking away from his interest in a valuable business." The music exec couldn't understand the illogical business decisions Thorpe had made, like opting to leave Robot Wars instead of remaining a minority partner.

Chandler asked Plotnicki if perhaps he had alienated the robot makers himself. "Did you sue any of them?"

Plotnicki said that he had sued Gary Cline and Carlo Bertocchini, in each case trying to defend the Robot Wars trademark.

"Don't you think those lawsuits had something to do with your reputation?" Chandler asked.

"No, I don't."

Chandler returned to the issue of the alleged conspiracy. Plotnicki made it clear he thought that all the phone calls between Thorpe and Roski "defied credibility." Of the entry forms Thorpe had lost from the 1997 Robot Wars, Plotnicki said, "I know he keeps everything. Just from my knowing him over the years, he's a stickler for keeping things. I find it incredible that he can't find them." Of Thorpe's awareness of Roski's new robotic combat event in 1999, Plotnicki said, "He might not have heard the word 'Battlebots,' but he certainly knew what Mr. Roski's plans or intentions were."

Plotnicki also referred to what he thought was Thorpe's tacit endorsement of Battlebots on the Delphi Web forum. Thorpe had written, "Trey is a dear friend and I wish him well with all he does, but I have no involvement at all with Battlebots."

Plotnicki called that statement "a high sign to the robot makers that Roski is okay, and they should compete in his event."

"But aren't they dear friends?" Chandler asked.

"I don't know," Plotnicki said. "It look's to me like a business partner, not a dear friend."

Three other witnesses took the stand that day as well. A communications specialist named Robert Wunderlich, hired by Profile, testified that the losses to Robot Wars in the wake of the cancellation of the U.S. event totaled $677,000. These were the damages directly attributed to the behavior of Thorpe, Wunderlich said.

Trey Roski took the stand as well. The CEO of Battlebots had been well-coached by his lawyers at Latham and Watkins and replied to every question with a crisp and succinct "yes" or "no."

Dave Chandler got right to the point in his direct examination of Roski. Was Thorpe involved with Battlebots, or were there any secret agreements between them? "No," Roski replied. Was Roski told by his lawyers not to talk to Thorpe about Battlebots? "Yes."

There was a short and insignificant cross-examination that failed to pierce Roski's armor of brevity.

Marc Thorpe was the third witness to take the stand that afternoon in Santa Rosa, and he testified again a month later in July. The second trial, in front of the same judge on the same issues, concerned Dave Chandler's demand that Plotnicki pay Thorpe the money due him under the 1999 settlement agreement. Plotnicki didn't even fly to Santa Rosa for that trial. Mistakenly, he figured it would be an uneventful rehash of the earlier proceeding.

On the witness stand, Thorpe appeared completely worn, reed-thin, his thick mane of black hair now speckled with gray. Behind his wire-rim spectacles, deep red circles arced under each eye.

Thorpe testified that he had not been paid the $250,000 from Profile Records, or received any royalties, despite the success of Robot Wars in Europe. He also claimed that after his controversial postings on the Delphi bulletin board, he had called some of the robot builders at home and sent them letters, urging them to attend the Robot Wars '99 event.

"Do you think that anything you could have said or didn't say to the robot contestants affected their decision whether or not to participate in Robot Wars '99?" Dave Chandler asked Thorpe.

"No, I don't. I think they had already made up their mind what they were going to do or not do."

"Mr. Thorpe, after March 1999, do you know if Mr. Plotnicki or Profile or Robot Wars LLC made any attempts to garner participants in Robot Wars '99?"

"I don't. Well, I mean, I know they sent out entry forms. So that's an attempt."

Bill Pascoe, the Marin-based counsel for Profile Records, cross-examined Thorpe. He asked Thorpe about his bankruptcy filing, and wondered whether it wasn't an effort to simply auction off the trademark to Trey Roski. (Thorpe denied that it was.) Pascoe also wondered if Thorpe had ever intended to rehabilitate Plotnicki's reputation.

"I didn't know what specifically I could do," Thorpe protested. "I vigorously encouraged Steve and Gary Pini to do positive things that I could report to the Robot Wars community, as an indication that the dark clouds had passed and that this was a new day."

Pascoe asked if it was true that Thorpe wasn't happy with his bankruptcy settlement. "I was happy that there was a resolution," Thorpe answered, "but I didn't think it was necessarily fair."

Isn't it true you told the robot makers that you weren't happy with the settlement? Pascoe asked.

"I expressed to people who talked to me my honest feelings. I felt like it was more of a surrender," Thorpe said.

Pascoe asked if communicating that unhappiness was consistent with Thorpe's agreement to promote the Robot Wars business. Thorpe responded, "I was acting consistently with the way I had acted before, so in that context I felt that, yes, it was consistent with what my idea of promotion was."

Pascoe wondered what the point was of the reference to Rodney King in the infamous March 13, 1999, post.

"Some humor," Thorpe answered. "I thought it would give some levity to a rather grim topic."

Pascoe read the entire e-mail aloud to the court and asked if it was consistent with "rehabilitating" the Robot Wars business.

"I was exhausted, frustrated, and frightened at my failing health," Thorpe said. "I was doing a lot of things to help the business. I was working on the press release. I was doing all I could, and that still was not enough. And then I was being accused of conspiring with Battlebots and not disclosing it before I knew about it, neither of which were true at all. Plus I couldn't get out of bankruptcy. That was what this was all about."

"We now call Stephen Felk to the stand."

In the week before the second trial, at Dave Chandler's urging, Marc Thorpe had called local builders and asked them to come to the courthouse. He told them only that they would be required to tell the truth about why they chose to attend Battlebots over Robot Wars in the spring of 1999. These were the key witnesses who could shed light on the question of Thorpe's ultimate liability. Had he killed Robot Wars in the United States? Or had Plotnicki fired the bullet himself? Nearly a dozen robot builders turned out for the trial. Felk was the first to walk to the front of the courtroom.

"Did you sign up for Robot Wars '99?" asked Chandler.

"No, I didn't."

"Why?" asked Chandler.

"I basically didn't want anything to do with Robot Wars."

"Did anything that Marc Thorpe said after February '99 have . . . anything to do with your decision not to participate?"

"No. I mean, Marc Thorpe had nothing to do with it. My decision not to participate was based mainly on Steve Plotnicki's behavior."

"And how were you aware of Mr. Plotnicki's behavior?"

"There was a lot of information on the [Delphi] forum, talking to other robot builders. The way that Robot Wars '97 was handled. There were a lot of things that contributed to my decision."

"Is there anything that Mr. Thorpe could have said to you but didn't say that would have influenced your decision?" Chandler asked.

"Sure, he could have told me that he had reached an agreement with Steve Plotnicki and Steve Plotnicki was no longer involved in Robot Wars. And if he would have told me that . . . if that had been the case, I would have begun to consider participating in Robot Wars."

"I see," Chandler said.

"When I first saw Robot Wars, one of the things I really liked about it was there was just this great sense of community. The appeal was to build the robots and then it was just hanging in the pits and seeing all the guys and all the builders and the camaraderie. . . . It was a great thing. I really wanted to participate in that. And I think that Mr. Plotnicki's behavior was just completely counter to the whole thing. . . . Here I am testifying in court. That was something I didn't want anything to do with."

"Why did you think you might end up in court?" Chandler asked.

"It seemed that Mr. Plotnicki had a tendency to sue people."

"Were you aware that he sued any other robot contestants?"

"I had heard a rumor that he had sued Carlo Bertocchini. And there had been another event called Robotica, and I believe he sued with that."

Profile attorney Bill Pascoe declined to conduct a cross-examination.

Grant Imahara, an ILM special-effects designer, followed Felk to the stand. Imahara's Deadblow was the second-season middleweight champion and sported a lightning-quick pneumatic axe, capable of hammering away at an opponent like a woodpecker.

"Why didn't you sign up for Robot Wars '99?" Chandler asked.

"I had heard that Profile sued Carlo [Bertocchini], one of the robot builders, and I didn't like that," Imahara said. "To have the company sponsoring the event attack one of the competitors in such a way, I felt that easily could be any of us, and I didn't want to be involved with that."

"Could anything that Marc Thorpe have said to you persuaded you to participate in Robot Wars '99?"

"No," Imahara said. "By the time Profile had damaged the event, Robot Wars, so [much] that I don't think that anything that Marc Thorpe could have done to convince me would get me to compete."

Bill Pascoe cross-examined Imahara and wondered if he hadn't formed this view through messages on the various Web forums—messages posted by builders who were in close touch with Thorpe, and who were hearing a biased interpretation of events. Imahara said he barely read the Web forum.

Pascoe then appealed to the judge, questioning the relevance of the builder's testimony.

"Doesn't it go to the very issue that I stated before I wasn't sure about," Jaroslovsky responded, "the extent to which Profile shot itself in the foot, as opposed to [Mr. Thorpe causing the] damages?"

Pascoe conceded the point and sat down. Chandler called Alex Rose, a member of the "Toro" team, to the stand. As the director of a nonprofit organization called the Long Now Foundation, the 30-year-old Rose was working on the construction of a 10,000-year clock—a mechanical timepiece robust enough to record the exact time for a hundred centuries. Upon completion, it is to be installed next to the Great Basin National Park in the eastern Nevada mountains. Rose was a quiet and intense member of the robotic combat fraternity.

"Did you sign up to participate in Robot Wars '99?" Chandler asked.

"I did not," Rose answered.

"Why?"

"Because, for the event [Robotica] in 1998, myself and my three partners probably put in, I don't know, a few hundred hours trying to build our robot, only to have it canceled by Profile's lawsuit. And we felt horribly betrayed by that and we knew that other people had already bought tickets to that event from far away, airplane tickets they couldn't redeem. So when they decided to have another event, it just seemed completely irrelevant to us that they would even ask us for anything."

"Did Marc Thorpe ask you to participate in Robot Wars '99?" Chandler asked.

"I believe he did call me, actually. And that was the only thing that started to make me think that things were starting to be okay with him. But when I found out that Profile was really still involved, it seemed irrelevant. And there was another competition, so why do it?"

"Did you read any electronic messages in or around March of 1999 from Marc Thorpe that influenced your decision?" Chandler was referring the "Rodney Thorpe" posting.

"No. I may have read those, but they certainly didn't influence my decision," Rose answered.

"Were you aware of the litigation between Mr. Thorpe and Profile?"

"Peripherally, yes."

"And you're aware that Mr. Carlo Bertocchini had been sued by Profile?"

"I had heard that rumor. I didn't have any confirmation."

"Did that have anything to do with your opinion of Profile?"

"All of the lawsuits had contributed to my negative opinion toward Profile," Rose said. "The amount of work [in 1998] that we went through only to have it canceled, and then somebody stepped forward with their own money to put on a new kind of event that was a private party [Robotica]. And then they sued him a week before the event so he couldn't fly to New York to defend himself. . . . And this lawsuit was then tried, it's my understanding, against Battlebots. At least they had the money to get lawyers out there and stop it. So it was a completely frivolous act that stopped someone who was really trying to do good for the robot community, where all these organizations are pooling their free labor. I thought that was deplorable."

Rose was cross-examined by Bill Pascoe. The Profile attorney wondered if Rose knew that Profile had tried to put on a Robot Wars '98 and asked for Thorpe's cooperation on it. "Would that have changed your opinion if you were aware of that?"

"The event being canceled proved to me that [one of the] parties didn't agree," Rose said. "The affected people [were the ones] basically providing the free talent. So it was irrelevant to me where the disagreement came from. It's irrelevant to me whether or not it was Marc or if it was Profile. The point was that the event was canceled due to something that was beyond the control of the contestants, who were actually the ones providing the real work behind the event."

Other builders also took the stand that day. Fon Davis, an ILM model maker who competed with the pink, domed flipping lightweight, Mouser MechaCatbot, testified that an informal boycott against Profile formed among the contestants, both online and in private communication. "We just kind of all agreed that the activity that surrounded Robot Wars was taking the fun out of the event," he said. "Attacking the contestants is definitely the wrong thing to do if you want participation."

Gage Cauchois mystified the lawyers and judge by stating that he didn't sign up for the 1999 Robot Wars because he had refurbished Vlad the Impaler for Battlebots. Afterward, he said, it exceeded the Robot Wars weight limit.

Ray Scully, a Battlebots participant who helped build the Robotica arena in '98, testified that Profile's legal maneuverings were "unsupportable." Scott LaValley, who had been there from the beginning and appeared on *The Tonight Show,* nervously opined on the witness stand that Plotnicki was "a bad seed."

Jim Smentkowski also testified: Plotnicki viewed the Marin County–based creator of Nightmare as a central figure in the conspiracy, as Smentkowski had been in close touch with Thorpe throughout the ordeal, and had posted one-sided updates about the dispute on his website, robotcombat.com. In this way, Plotnicki believed, a skewed version of events had filtered throughout the community and poisoned the well for the Robot Wars business.

In his testimony, Smentkowksi described the time and hours he had spent preparing for the '98 event and said he had felt betrayed by its cancellation. In cross-examination, Bill Pascoe read aloud something Smentkowksi had said during a deposition several months before: "It was stated several times publicly on the forum that if Marc would have been involved, [the builders] would definitely go to Robot Wars '99." That statement was the heart of Profile's case: that Thorpe could have gotten the builders back to Robot Wars if he wanted to, despite all those lawsuits in the past. Smentkowski admitted he had said that during his deposition.

Chandler stood up for redirect and tried to minimize the damage.

"Mr. Smentkowski, after listening to your fellow robot contestants this morning, do you believe that anything that Marc Thorpe could have done or could have said would have made them participate in Robot Wars '99?"

"Not as long as Profile was involved," Smentkowski said.

"Prior to giving your testimony September 2, 1999, had you heard any of your fellow contestants make statements such as they made this morning?"

"Oh, yes," the maker of Nightmare replied.

In their closing argument, the two lawyers painted very different portraits of the demise of the original Robot Wars in the United States. Dave Chandler claimed that the Robot Wars partnership faltered when Steve Plotnicki's attention turned from the lawsuit with Tommy Boy Records and its parent, Warner Records, that stemmed from his separation with partner Cory Robbins. Chandler conceded that Thorpe's e-mails were not in the spirit of the agreement, but insisted that the artist used his "reasonable best efforts"—as the agreement stipulated—to rehabilitate Plotnicki's reputation.

Chandler assumed for the sake of argument that Thorpe's postings were a breach of contract. Still, he said, the builders' testimony proved that Plotnicki caused their boycott of Robot Wars, not Thorpe. Plotnicki's attorneys, he said, didn't prove that Thorpe's lukewarm statements directly proved the damages.

"[Marc Thorpe] called these guys up. He went to Battlebots. He tried to get them to sign, and they wouldn't sign. They were mad. We learned today why these people wouldn't do anything with Robot Wars. There was a boycott. And that boycott was a direct and proximate result of Profile's conduct and Profile's conduct alone."

Chandler said it didn't make a difference whether the lawsuits against Carlo Bertocchini or Gary Cline were technically justified. "It's the perception among the talent pool for Robot Wars," he said. "These people are the people that make these things that make the money for Robot Wars. And by suing these people, making these people fearful of you, they don't want to play with you. It's a game to them. It's entertainment, it's community, it's friendship. It's all those things, like baseball games or something.

"And Profile made it not fun by this cloud of litigation that they placed over their heads. . . . [Profile] made its own bed."

Chandler concluded, "I submit, Your Honor, that even if the court finds there was a breach and even if the court finds that they've got the right party, they haven't been damaged because of anything Marc Thorpe did. And there is virtually no evidence to the contrary."

Profile lawyer Bill Pascoe began his closing argument by hammering Thorpe for failing to live up to his end of the settlement agreement. Profile had bargained for Thorpe to do everything in his power to promote and rehabilitate Robot Wars. "The sins of Profile, if any, were in the past . . . and we believe that the actions that Profile took, it took on the advice of counsel to protect the trademark."

"Well, now wait a minute," Judge Jaroslovsky interjected. "Mr. Plotnicki is a businessman. There is no question that there may have been advice to counsel on a legal right to sue. On the other hand, from a business standpoint, it looked to me before and still looks to me like it was the stupidest thing that Mr. Plotnicki could have done."

Pascoe argued that robot makers were alienated not by Plotnicki's legal actions, but by the negative way they were represented on the Web forum, by builders like Jim Smentkowski, who was getting a distorted version of events directly from Thorpe.

If Profile didn't sue Carlo Bertocchini and Gary Cline, Pascoe argued, it "would have lost the Robot Wars name. . . . There was a duty on Profile's part to protect the entire joint venture." Indeed, he said, Robot Wars was little else besides a trademark and a set of copy-written rules, which had to be zealously guarded.

Pascoe insisted that it was Thorpe who turned the robot makers against Robot Wars. First, he tried to hold the 1997 event when Profile refused to put it on. When Profile sued to stop the event from happening, Thorpe invited all the robot makers to send letters to the judge, in an effort to demonstrate the harm that canceling the event would cause. By the way, Pascoe noted, the flood of letters that resulted from that plea revealed Thorpe's ability to direct the robotic combat community in causes he truly supported.

But, Jaroslovsky asked, didn't all that happen before Profile sued the two robot builders? How do we know the community would still be so compliant after the legal turmoil had commenced?

"That raises a good point, Your Honor, the chronology of events. What was really done and what wasn't." Profile's attorney then reviewed the series of negative posts by Thorpe and by builders close to Thorpe on the Web forum. "They start to pick up, I think correctly, that Marc is really not too enthusiastic about this whole event."

In the first 10 days after the settlement, Thorpe violated it, Pascoe argued. "That comes with a price, unfortunately, to everyone."

Rather than get a judgment against the broke Thorpe family, Pascoe asked the judge to revise the settlement agreement and throw out the royalties due Thorpe. The artist hadn't earned it, he said.

Judge Jaroslovsky made a few comments to the parties before he adjourned the proceeding and reflected on a final decision. While he said he felt Thorpe was an honest man, he noted that the California artist "seems to have never been able to accept the fact that when you sell your soul to the devil, you can't pretend it's still yours."

Of Plotnicki, Jaroslovsky said he "utterly failed to realize, as far as I can see, the personal devotion that the participants have to Mr. Thorpe, and the difficulty of cooperating where there are two styles that are so divergent."

He added, "Frankly, it seems to me, Mr. Plotnicki, that you shot yourself in the foot right from the get-go by getting involved in litigation."

With that, he left the courtroom, and took the matter under submission.

The parties waited two weeks for a decision, and there was more than enough time to think about the past.

Marc Thorpe had more than his normal share of regrets. He regretted no longer being the central figure in the now-successful, televised sport of robotic combat. He regretted going into business with Profile Records in the first place, and being blind to how cruel people could be to each other in the real world of business and entertainment. He really lamented all the time he had lost with his teenage daughter Megan, and the pressures she had endured growing up under the weight of her father's legal problems. Finally, he regretted what had happened to his marriage.

Somewhere in the years-long haze of court appearances, mounting debt, and the humiliation of their family records being laid open for public viewing, Marc and Dennie Thorpe's already troubled marriage permanently fractured. Thorpe moved out of his house to his next-door workshop but couldn't afford another place with his current financial problems. He didn't particularly want to talk about it. "She's really suffered as a result of all this," he would say. "There's been a huge amount of frustration, anger, and fear. It was an awful and oppressive burden."

Although Thorpe doesn't blame Plotnicki for the breakup of his marriage, he says, "He knows all this legal pressure has a dramatic effect on families."

Thorpe was full of regrets, but he never regretted fighting back and holding his ground. "I don't want to be characterized as a victim," he said. "If I had given in to what Steve wanted, my financial situation might be better. But I surely would have had the same stresses and strains from being under his control. I know how well I do being my own person, and how poorly I do feeling owned by someone."

For his part, Plotnicki had few regrets. Back in late '98, he had sold his rap label to Arista Records. *Billboard* magazine pegged the sale at $12 million and characterized the negotiations as long and drawn-out. Afterward, Plotnicki was effusive, telling *Billboard,* "It's personally more rewarding to be bought out by a great label, as opposed to just someone with a lot of money."

As a result of the transaction, Joseph "Run" Simmons was transferred from Profile to Arista. The original rapper and one of Plotnicki's old foes took the opportunity to fire a parting shot: "I feel like

I went from the pit to the palace, from the very bottom to the very top," Simmons told *Billboard*.

Plotnicki kept his dance and jazz labels, plus his music publishing company—which meant he would still earn royalties on all the early rap hits Profile had signed. As a hobby, he sold olive oil from his Manhattan apartment. "It's a goof pleasure to me," he said. On occasion, Plotnicki agreed to put himself on the couch, to speculate about the eerily similar control battles—with Run-D.M.C., Cory Robbins, and Marc Thorpe—he had waged over the course of his career as an investor and music entrepreneur. In each dispute, he ended up feuding with larger, more powerful figures—Russell Simmons, Time Warner, and the Roski family, respectively.

"I guess it is weird," he acknowledged. "My wife says it's because I'm unyielding. I always credit it with being the son of a Holocaust survivor. I don't like to give up what's really mine. It makes me dig in."

That August, Judge Jaroslovsky issued a written decision worthy of King Solomon.

"The principal of Profile is Steven Plotnicki, who appears to be almost a caricature of an East Coast entrepreneur: brash, aggressive, combative, and litigious," the judge wrote. "Thorpe appears to be his exact opposite: West Coast, reserved, quiet, personable, naïve. With the benefit of hindsight, a 50–50 partnership between the two of them had no chance of succeeding.

"The essence of whatever success Robot Wars has enjoyed is a small group of dedicated robot builders anxious to compete against each other. Thanks to the Internet, Thorpe was able to build a very personal relationship with them. This personal relationship, and Thorpe's inability to make the builders comfortable with Profile, has resulted in the impasses now before the court."

Jaroslovsky agreed Thorpe had broken the settlement with his lukewarm Web postings. "His agreement did not call for him to be honest with the participants, it called for him to bring them back into the Robot Wars fold. If he could not do this in good faith, he should not have made the agreement. He was not free to take Profile's money and then tell the participants that Profile made him do it."

But the judge said that fixing damages in the case was difficult. He dismissed as flawed the testimony of Profile's expert, Robert Wunderlich, and wrote, "By bringing suit against builders, Profile damaged itself far beyond Thorpe's ability to smooth things over, even if he had done his best. The court perceives that Profile did not take into account the changing times in assessing the value of Thorpe's attempts to rehabilitate it. Even as recently as two or three years earlier, it is very possible that Thorpe's endorsement would have brought most or all of the participants back into the fold. However, the information explosion of the late nineties changed many things, especially among tech-savvy robot builders who are more likely than the general population to take full advantage of e-mail and the Internet. Thanks to these developments, the builders across the world were in constant, instant communication with each other. Information was shared to the extent that a strong anti-Profile sentiment developed in a significant portion of the builder community. Given the extent of their knowledge of the circumstances, it is not surprising that they would correctly view anything Thorpe said or might have said as a grudging consideration for a reluctant compromise."

For this reason, Jaroslovsky preserved Thorpe's right to a royalty from Robot Wars, but then attempted to award damages to Profile in proportion to the degree that Thorpe's Web postings had caused Robot Wars harm. The judge calculated Thorpe could have lessened the negative feelings toward Profile by one-third. So, he awarded Profile one-third of its estimated $677,000 in damages—or roughly $225,000. He subtracted that number from the $250,000 Profile already owed Thorpe under the settlement, and ordered Plotnicki to pay Thorpe the remaining $24,000, in addition to the 10 percent royalty.

Both parties had something to be upset about. The big cash award that could resolve Thorpe's many debts wasn't going to materialize any time soon. On the other hand, the judge had found that there wasn't sufficient evidence of a secret conspiracy between Marc

Thorpe, Trey Roski, and Bob Leppo, and had maintained Thorpe's royalty—10 percent of all the income Plotnicki made off licensing Robot Wars.

But it wasn't over yet. Both sides appealed the ruling to the district court in San Francisco. In the meantime, Steve Plotnicki wasn't going to pay Marc Thorpe a dime.

Chapter 10

The Final Showdown

If you think our robots are tough, wait until you see our lawyers.

—2001 Battlebots advertisement in toy licensing magazines, warning off the industry's infamous copycatters

In the year 2001, fresh from their journey to exalted levels of media recognition, the U.S. robot builders began to take quite seriously their mutual endeavor. What began as a hobby in the early nineties had become more professional. Christian Carlberg left his job at Disney, where he designed theme park rides; Gage Cauchois stopped building $500 art deco lamps from his East Bay studio; Carlo Bertocchini quit his job in Silicon Valley designing plastic injection molds. They all were going to devote themselves full-time to robotic combat.

However, it couldn't support them completely. Bertocchini supplemented his income by selling robot books and kits from his website, robotbooks.com. But his focus remained on the combat, the all-consuming, wife-frustrating obsession. "There's nothing better than being able to quit your job and do your hobby for a living," he said. "My friends are jealous."

He could afford to make them jealous because each time Comedy Central broadcast a Biohazard fight, he got a $1,700 royalty check.

New Battlebots toys were coming out that summer, and each time someone purchased a $60 remote-control Biohazard model, Bertocchini would get a 15 percent cut of Battlebots' 15 percent royalty—or about $1.35.

For the most serious builders, this was only the beginning: Sponsors were also entering into the equation. During the early years of robotic combat, sponsorship meant steep discounts on robot parts. Builders would, for example, buy a $700 speed controller from L.A.-based manufacturer Vantec for the discounted rate of $200, in exchange for slapping a Vantec sticker on the robot.

With the sudden prominence of *Battlebots,* there were now greater possibilities: The builders could sell companies on the prospect of cheap TV exposure to millions of fans, and a core audience of eager, 18-to-49-year-old mayhem-loving gearheads. The BBC, which considered Robot Wars "sports entertainment," didn't allow sponsorships, but Comedy Central treated it as a sport (the ironic sensibility notwithstanding) and had no objections to sponsorship logos floating across the screen on the mechanical gladiators.

Gage Cauchois went to the Oakland library and researched the whole alien concept of sponsorship in his usual meticulous, bordering-on-obsessive way, sketching out a two-page proposal that he sent to 25 companies. He didn't hear back from Radio Shack or Compaq, but he did hear from Loctite, a German company that sold industrial parts and adhesives. Loctite agreed to pay $25,000 for the construction of Cauchois's new super heavyweight, Vladiator—and even paid extra to show off the robot at trade shows. Vladiator was a fast, powerful four-wheel rectangular bot with an articulating spear on its front end. It featured the Loctite logo in bright-red letters on each side of its frame.

Christian Carlberg didn't even have to write a proposal. He got an unsolicited call from a marketing executive at Wizards of the Coast, the company that makes the Dungeons and Dragons–like card game, Magic: The Gathering. "My worst fear was having a tobacco company call me," a relieved Carlberg said. The game company agreed to finance the construction of a new super heavyweight, called Dreadnought. Over the first few months of 2001, Wizards of the Coast paid 60 grand in parts and hourly wages for Carlberg and his two team members, Brian Roe and Luke Khanlian. It was going to be their dream robot—no expense would be spared. The finished super heavyweight weighed 310 pounds and sported two five-inch spinning discs on the back of a ramp-shaped base. A huge Magic: The Gathering illustration swathed the entire facade of titanium armor.

"They have a lot of money invested in this," Carlberg said before the upcoming competition that May. "We are both invested in seeing it do well."

That sentiment was repeated across the upper echelons of the builder community. Carlo Bertocchini signed with the Long Island–based industrial parts manufacturer Berg. Donald Hutson of Tazbot shook hands with execs at Cincinnati Machine Tools. Alex Rose and Reason Bradley, the makers of Toro, struck a deal with Internet messaging company ICQ. At the Battlebots event in May 2001, these robots and many others were upgraded with sponsor money and festooned with sponsor stickers. The builders were beginning to consider robotic combat a business—as the seed that might sprout into something like NASCAR.

The previous fall, a few months after the TV series started its run, 150 robots entered the Las Vegas tournament. After nearly a year of regular TV coverage, plus the added exposure of *The Tonight Show,* more than 400 teams signed up for the first event of 2001.

There were wedges and flippers, hammerbots and spinners of every kind, and Trey Roski was going to let them all enter the tournament. "Look at what the sport did for me," was how he explained his expensive decision to exclude no one. Battlebots put the 300 rookie robots into preliminary rounds, to reduce their number to the few dozen that would join the veteran robots in the televised competition over the long Memorial Day weekend. The preliminaries were conducted without arena hazards, to better preserve the robots that actually made it. And to discourage nonserious builders from acquiring a free pass to the tournament by signing up with a robot that didn't exist, Battlebots raised entrance fees from $100 to $250 for unsponsored newcomers (bots with sponsors paid a little more).

Roski and Munson also moved the event to a larger venue on Treasure Island, the wind-buffeted former naval base in the middle of the San Francisco Bay, accessible between the two spans of the Bay Bridge. In the new setup, the Battlebox arena and the pit area occupied side-by-side hangars. The pit area was a vision to behold: Here were teams from around the country, bunched shoulder to shoulder, their equipment and robots spread out on row after row of wooden tables. A computer-assisted-design (CAD) company occupied a table off to the side, offering the competitors who made the final rounds thousands of dollars worth of free software in exchange for putting a sponsor sticker on their robots. Camera crews from Comedy Central and other media outlets roamed the hall. There were 400 robots and their determined creators in there, and the sound in the hangar—well, how do you describe the low drone of 1,000 gearheads mumbling to themselves about insufficient armor?

On the first day of safety inspections, Roski addressed the crowd from a balcony above the pit floor. "I don't feel so crazy anymore," he said. "There are a lot of other nuts in this world, and they're all here." The builders responded with thunderous applause.

Compared to the first generation of prototypes back in 1994, these new robots had supernatural abilities. Jascha Little and his dad, Scott Little, two unusually tall mechanical engineers from Austin, Texas, spent six months designing a hammerbot called The Judge. The robot was sheathed in black Kevlar fabric, lending it the appearance of a ghoulish medieval executioner. A pneumatic hammer provided the offense; it started horizontally, with its spiked tip pointed up, and rotated 180 degrees with such force that it put nickel-size holes in the metal floor. In its first match, The Judge repeatedly speared Fuzzy Mauldin's orange plow Iceberg so gruesomely that it brought the crowd roaring to its feet in a half-minute-long standing ovation. Afterward, tournament officials asked the Littles to remove the spike from the end of The Judge's hammer and use the blunt edge instead, so as not to damage the floor. At the end of the weekend, father and son were awarded the prize for best-engineered robot.

Sausalito high-school friends Alex Rose and Reason Bradley, makers of the super heavyweight Toro, also pushed the envelope. They were secretive about the customized pneumatic technology that powered Toro's flipper, and now they applied the same system to a compact, dual-wheeled middleweight called T-Minus. In its first match during the preliminary rounds, against the mace-swinging Halo, T-Minus collided with its opponent and flopped over onto its back; the crowd wondered if the rookie's career had ended prematurely. But then Rose activated T-Minus's pneumatic flipper, and in a sudden burst of CO_2 gas, T-Minus sprang into the air, executed a two-and-a-half-turn somersault, and landed upright. Nobody had ever seen a robot do anything so . . . athletic. Members of the audience jumped back to their feet. T-Minus lost in the middleweight quarterfinals, but returned to capture the middleweight rumble and the imagination of the fans.

If both these teams raised the bar, then Terry Ewert and Team Whyachi broke it in half and threw it out altogether. Ewert, the owner of Westar, a Wisconsin company that made meat-processing machines, was the designer of spinning robots Whyachi and Son of Whyachi. Back home in his basement in Dorchester, Wisconsin, he kept a hovercraft, a wind tunnel, a remote-controlled helium blimp, and a trebuchet—a medieval catapult made of stainless steel. It stood two feet tall and chucked baseballs 50 feet. All those contraptions had won either first or second prize at local elementary-school science fairs. Ewert's four young boys, aged 8, 9, 12, and 13, helped too.

Ewert and his crew were the talk of the pits that week. Their two walking helicopter-spinners looked slick and expensive, and the robots' color scheme matched the team's red-and-black-striped NASCAR racing uniforms (with the black stripe covering the stomach—an effective strategy for concealing a little paunch, Ewert pointed out). Word also got out that Ewert had run his bot-building effort through the Westar machine shop. Nine employees spent six weeks building the robots, sparing no expense. It seemed to confirm the worst fears of the builders who weren't running out and getting big-money sponsorships: The sport was going big-time, and that the garage gearheads wouldn't be able to keep up.

Other competitors also complained that Team Whyachi was playing fast and loose with the rules. Before the competition, Ewert had spent weeks poring over the Battlebots guidelines and noticed that walking robots—called stompbots—were allowed 50 percent more weight in their class. The rule dated back to Marc Thorpe's Robot Wars: Thorpe had wanted to encourage more walking robots to add variety, and had in mind walkers like Mechadon, built with actuators that moved the legs back and forth, the way insects walk.

Terry Ewert took another approach. He essentially hacked the rules. His robots were wheeled, but the wheels were attached to camshafts, mechanisms that convert circular motion into linear, back-and-forth movement. Connected to the cam setup on each side were eight rectangular legs, which shuffled along in a trapezoidal pattern, one set of four always gripping the floor.

Ewert and employee Dale Hammel named their robots after a bit of Westar jargon. "Whyachi" was code for any maneuver that harshly eliminated an opponent during the card games they played in the break room. So the super heavyweight, weighing in at 488 pounds, was dubbed "Whyachi." The 315-pound heavyweight, a scaled-down duplicate of the first design, became "Son of Whyachi." Adding up parts and man-hours, Terry Ewert had invested nearly $130,000 in the robots. His investment was rewarded that first day of the tournament by the awe of his fellow robot builders, who wandered over in droves to check out these unique and fearsome machines.

Sitting in the bleachers later that day, watching the enthusiasm of the crowd and the passion of his fellow gearheads, Ewert felt as if he had come home. "Yup, we belong here," he told his wife, Lisa. "This is what we were meant to be doing."

But after his first fight, Ewert got an unwelcome surprise. Son of Whyachi was facing Shaka, a four-wheeled robot with a lifting plow on its front end, designed by mechanical engineers from Southern California. Shaka took a few good hits from Son of Whyachi's helicopter rotor, and then Son of Whyachi took a big blow from Shaka, and the helicopter weapon disengaged. It became a pushing match, each bot trying to shove the other into the spikes and other hazards of the arena. After three minutes, the fight went to the panel of three judges, who narrowly decided in favor of Son of Whyachi.

The audience booed. The crowd was rooting *against* Terry Ewert's team.

Dale Hammel leaned over and said to Ewert, "When you get booed, it means you're doing good and winning." Hammel had seen NASCAR's Jeff Gordon race at the Ford City 500 at the Bristol Motor Speedway in Tennessee earlier that year. The crowd at these events always rooted against the stock car racing champ. It was a good sign, Gordon always said, it meant they were scared of him.

But then Lisa started wandering the pits between matches, and she started hearing things, disturbing things, things that she hoped 13-year-old Jake, who had traveled with the team, wouldn't hear. Opinion was against Team Whyachi and it was personal. Their robots looked *too* good. Those NASCAR pit uniforms seemed *too* expensive. Word was getting around: They were an actual company, not the usual garage junkies who built and fought robots as a hobby. They had their own drill press, lathe, and—even more exotic mechanical toys? They had nine employees working full-time over four weeks? "I don't have the use of a mill and lathe," Lisa heard one guy say. "Those guys are ruining a family sport."

Ewert said it didn't bug him. But then his super heavyweight, Whyachi, faced another spinner called Odin 2. Ewert thought he had another easy victory, but in the Battlebox, Whyachi took a big hit and its weapon motor disengaged. Odin 2 also lost its weapon, and it became another pushing match. But this time, at the end of three minutes, Ewert, Dale Hammel, and Ewert's oldest son, Jake, went into the arena and stood next to the referee, with their rivals on the other side. In front of the cameras and a raucous, yelling crowd, the ring announcer declared that the victory went to Odin 2, 28 to 17.

The crowd went wild. The veteran builders in the crowd jumped to their feet and stomped on the bleachers. The corporate juggernaut had been stopped! Down with Team Whyachi!

Back in the pits, pacing up and down the crowded row of wooden tables, Ewert was nearly inconsolable, ranting to anyone who would listen. "The judge's decision was wrong. Whyachi got in more blows. It caused more damage. How did they calculate that lopsided point total? And what's with that booing? I'm here with my son and my wife. This is supposed to be a family affair . . ."

Ewert was hurt. And disappointed. Whyachi was his first robochild and best prospect. Son of Whyachi, the heavyweight, the clone, had almost been an afterthought.

That week in the pits, Christian Carlberg was the gearhead equivalent of Michael Jordan hawking his Nike shoes on the court. His team wore black short-sleeved Oxford shirts, with Magic:The Gathering logos on the front and full Magic illustrations splashed over their backs. A Wizards of the Coast publicity rep followed Carlberg around, distributing press releases announcing the sponsorship. Carlberg even evangelized for the card game. "It's an intelligent competition, just like robotic combat," he told reporters. "There's tons of overlap. Everybody here plays Magic."

Actually, it was hard to imagine that any of the competitors played Magic. There was no time, in between the robot building, the robot fighting, and the fending off of the pesky girlfriends who were annoyed about the robots. Still, Carlberg was bringing in sponsorship money from a company not directly involved in industrial parts. Naturally, Wizards of the Coast planned on seeing its sponsorbot on television, in front of those millions of mechanical obsessives and their kids. They had paid good money for the exposure. But like all the other rookie robots, Dreadnought had to fight its way through the preliminary rounds. It would have to survive at least two matches to reach the realm of the television cameras, tuxedoed ring announcers, and buxom ex-*Baywatch* starlets.

Dreadnought got a first-round bye, and in the second round fought against a relatively unsophisticated, gas-engine spinner called O.J., slapped together by engineering students at Florida Atlantic University with an $8,000 donation from the school.

It was David vs. Goliath, a sponsorbot versus a so-called ghettobot—a robot without the benefit of patrons or expensive parts. But as Carlberg knew, in this evolving sport, anything could happen. Seven years of competition had highlighted a strong element of luck and unpredictability that was difficult to surmount with money or preparation. A robot could rush at an opponent, bungle into the spike strip along the sides of the arena and expire in a smoky, embarrassing mess. Even worse, it could sit inertly in its starting box, impervious to all radio commands sent from ringside. Perhaps some obscure component had burned out during a test before the match. Perhaps there was no explicable cause: Complex machines were sometimes erratic.

Stephen Felk liked to characterize this uncertainty in terms of the will of the "robot gods." Only a divine and sometimes malevolent force from above could account for the element of randomness that sometimes afflicted these competitions.

Unfortunately for Carlberg and Dreadnought, the robot gods were vindictive that day. Charging out of its box, Dreadnought flew across the arena and struck its opponent with the spinning blades on its rear. On impact, one of the discs sucked a loose piece of metal into its whirling maw, and the blades stopped spinning. O.J. also stopped rotating, and it became another of those pushing matches, back and forth over the killsaws and around the arena, like two boxers in an exhausted clinch. Dreadnought seemed to gain a slight advantage, and at the end, O.J. stopped moving. But before the ringside referee could begin a knockout count, the buzzer sounded. The match was over, and the decision went to the panel of three judges at ringside, each wearing an official-looking red sport coat.

The competitors felt these decisions added a further flavor of unpredictability to the tournament. One judge was the head of the local robot hobbyist club, another handed out a business card that proclaimed her "the world's first robot psychiatrist," and another had gone to middle school with tournament organizer Greg Munson. They graded the combatants on a point scale of one to five in damage, strategy, and aggression. The last two categories seemed particularly subjective, and many participants felt they were inconsistent with the exacting, scientific nature of the sport.

So perhaps it wasn't surprising that the three-judge panel passed down an eight-to-seven decision—for O.J.

Carlberg couldn't believe it. He bolted around the corner of the Battlebox and darted over to the judges. No longer the affable Michael Jordan, he was now a fuming, agitated John McEnroe. "[O.J.] did not survive the match!" he protested. "I damaged him! He did not survive the match!" Instead of a tennis racquet, Carlberg held his radio-controller, and he violently jammed the antenna up and down with his other hand.

The judges stood up and listened to Carlberg's frenetic objections. "O.J. was moving on his own power at the end," one judge said. "He escaped and moved away under his own power."

"I'm sorry," said Carlberg in disgust. "I'll tell you, I don't think this is the right call."

A crowd formed around them. Munson stood nearby and meekly supported his friend Carlberg's case while trying to avoid the appearance of intrusion. The argument went back and forth, hinging on the subtlest particulars of an unremarkable contest.

"You're arguing an eight-to-seven decision," one of the judges finally told a distraught Carlberg. "It can go either way. We're looking at the whole match, not the last few seconds. In my opinion, you lost. By a fraction, but you lost. You could have knocked him out. I saw that guy still moving at the end. But you backed off. You thought he was dead, you played your hand, and you backed off. You lost."

Carlberg scoffed at the judges and walked away in disbelief. "I don't think that I made any friends back there," he said later.

His team brought three other bots to the event, each adorned in Magic illustrations. They all made the television rounds. But there was no mistaking Carlberg's disappointment that day: His $60,000 sponsorbot was dead.

The next day, Terry Ewert's last hope, Son of Whyachi, began winning its way through the television rounds. The heavyweight spinner destroyed Jim Smentkowski's Nightmare in 48 seconds, snapping its gearbox in half and splashing gear oil over the Battlebox floor. Then it knocked out another spinner, Mechavore, before facing the 210-pound forklift wedge Hexadecimator in the semifinals. Its Seattle-based builders, including Tim Patterson, the creator of the DOS operating system, added extra armor to protect against the whirring helicopter rotor, but even that upgrade didn't matter: The forklift never got close enough to inflict any damage on Son of Whyachi without getting slammed by those rotating hammers. The bout lasted two minutes.

Biohazard, meanwhile, was waiting in the finals. Sleek, indestructible Biohazard, the devil's spawn. Impossibly low to the ground, an innocuous pizza-box on wheels until that arm popped up and its enemy was on its back, like a beetle, wheels futilely spinning in the air. In five years, it had won 31 matches and lost twice, once to Stephen Felk's Voltarc, once to Gage Cauchois's Vlad the Impaler.

It was the match everyone would remember: Bertocchini, the champion, wearing his team uniform, hospital scrubs, against the financial and technological might of Terry Ewert and Team Whyachi.

Biohazard in the red corner; Son of Whyachi in the blue corner. The rolling Christmas tree lights danced from red to green, inviting the low drone of the buzzer.

The battle lasted the entire three minutes. Halfway through, it seemed Carlo Bertocchini would win another title. Son of Whyachi was taking a terrible beating under Pete Lambertson's hammers and one of its bracing rods had come loose and flared wildly about, disabling the helicopter weapon. Moreover, Biohazard was strategically positioned to block all avenues of escape. Ewert was desperately trying to restart the spinner, but without any traction on the debris-strewn floor, the robot's base was rotating instead.

But then, somehow, after all those devastating hits by the Pulverizer, the little Delrin feet of Son of Whyachi found their footing. The base stopped turning, and the helicopter rotor resumed its invisible blur. The crowd gasped: Son of Whyachi was back from the dead. Terry Ewert inched his robot out from under the hammer and into unthreatening territory.

Suddenly, Biohazard appeared inert. The moment before, it had been pushing its rival around. Now it just sat there, seemingly anesthetized to radio commands. Bertocchini looked down at his controller in horror, madly pushing the joystick back and forth, without effect.

Son of Whyachi inchwormed over toward Biohazard. It moved so slowly that there was more than enough time for crowd members to howl and clutch each other with anticipation. BLAM! Son of Whyachi reached Biohazard and connected. An entire panel of Biohazard's titanium armor was ripped from its frame—and reappeared across the arena, twisted and bent. Biohazard still wasn't moving. At ringside, the referee standing over Bertocchini started to count down from 10. The crowd screamed with excitement at the possibility of an upset.

A look of twisted despair passed onto Bertocchini's face. His champion robot seemed finished. Then, incredibly, it began moving again. Biohazard was receiving commands. For the second time in a minute, a robot had been resurrected. The crowd howled. There was half a minute left. Son of Whyachi inched back toward Biohazard, that loose bracing rod still madly whipping in the air. The two robots collided again, the force of the blow sending Son of Whyachi careening across the arena. Another panel of armor was ripped off Biohazard. Its electronics were now exposed. The crowd inhaled together.

Again, Biohazard wasn't moving, and this time, it seemed definitive. While he maintained a straight poker face, Bertocchini's shoulders sank. He gave the customary signal of surrender to the referee: "Tap out," he said. The ref started the countdown, 10, 9, 8. Son of Whyachi was still moving, and Terry Ewert couldn't believe it, he was about to knock out the champion and win the heavyweight title, and then—wait! wait! unbelievably, Biohazard started to react to radio commands again, moving slowly and almost imperceptibly toward the center of the ring.

Then the buzzer sounded. The ref had called for a knockout, even though there were 10 seconds left on the clock. He evidently hadn't seen Biohazard stir at the last moment and completed the countdown. The crowd began to murmur its disapproval. How could there be a knockout if both robots were still moving? Trey Roski rushed over to consult with the judges. Confusion engulfed Treasure Island.

For several long, incredulous minutes, the crowd didn't know who had won the championship match. Even though he had surrendered by declaring a tap out, Bertocchini was arguing that Biohazard had been moving at the end. But no one wanted to finish the last 10 seconds of the match, and there was little point, since the combatants were at opposite ends of the Battlebox and looked utterly depleted. Finally, Roski and the referees conceded Bertocchini's argument and decided to turn the matter over to the judges. There was no knockout; the outcome of the match would be decided on points. Another minute passed. Spectators whispered to each other about the incredible contest: both robots, reincarnated; the damage to the seemingly impervious Biohazard; the regenerative power of Son of Whyachi, after getting savaged by the Pulverizer and losing one of its helicopter braces.

The judges returned their verdict. They jotted it on a note card and handed to the tuxedoed ring announcer Mark Beiro, who returned to his spot at the center of the arena. Terry Ewert, Dale Hammel, Lisa Ewert, and young Jake stood on one side, while Bertocchini, his wife, and team member David Andres stood on the other. The crowd quieted.

"The judges have reached a decision. By a score of 29 to 16, the new heavyweight Battlebots champion is . . . Son of Whyachi!"

Terry Ewert and his crew threw up their arms and started to jump up and down in triumph. And the crowd erupted—reluctantly drawn around to cheer for Team Whyachi. Even the veterans had been won over by the ingenuity and power of the amazing walking robot. It had demonstrated its superiority in the evolutionary world of the Battlebox.

As for Bertocchini, he was stoic, though disappointed. "Losing is an inevitable part of the game," he said. However, he wouldn't lose to the same bot in the same way, ever again. He was heading home to upgrade Biohazard, to prepare for the next tournament that fall. *Adapt or die*—that was the Darwinian imperative, and the battle cry of his favorite sport.

Not all of the original members of the robotic combat community were along for the thrilling ride at the May 2001 Battlebots competition. Marc Thorpe wasn't on Treasure Island: still immersed in legal appeals to Judge Jaroslovsky's verdict in his bankruptcy case, he didn't dare associate himself with the rival to Robot Wars, lest he dig himself a deeper hole.

Dan Danknick wasn't there either. The Los Angeles–based robot builder and co-host of the 1999 pay-per-view event had fallen out of favor with Trey Roski. The reasons went back to the previous year, when Danknick discovered he had cancer.

He caught it early, and surgery saved him. But the experience sent tremors through his life. The maker of Agamemnon and War Machine reassessed everything, including his job as a computer programmer, which he believed might have generated the stress that led to his illness. Danknick quit his job and looked for new ways to make his passion—robot building—into a sustainable business.

His effort focused on the Team Delta website: Since '96, Danknick had designed specialized electronic robot parts, such as power supplies and speed controllers, primarily for his own robots. After his diagnosis, he expanded his offerings and aggressively pitched them to the entire community, hoping sales from the business could pay for at least half his monthly mortgage.

But he knew that to accomplish that, he would have to get involved in other sectors of the robot universe. Danknick had already staked out impartial ground in the fractured world of robot competitions. Back in '99, after fellow competitor and attorney Tony Buchignani told him he could be sued for his vituperative anti-Profile postings on the Internet, Danknick and Buchignani took neutral positions in the Thorpe-Plotnicki imbroglio. Unlike many of the other veterans who still bitterly hated Plotnicki and felt a strong loyalty to Roski, the members of Team Delta said they were willing to participate in Robot Wars.

After his cancer scare, Danknick decided to actively court the rival robot competition. He wrote a letter to the producers at Mentorn in the United Kingdom, complimenting them on their production of the MTV pilot. He also heard that a reality TV production company called Nash Entertainment was ramping up production of its own robot show, *Robotica* (scooping up the name that Gary Cline had abandoned after getting sued in 1998), and Danknick contacted them too. Whichever direction the phenomenon of robotic sport moved, he wanted to be there. It was good business.

Even if it was sure to infuriate his old friend Trey Roski, a few months before the Battlebots competition on Treasure Island, Viacom announced it was finally bringing Mentorn's version of Robot Wars to the United States. Fresh from dropping their FIRST project, Viacom execs Albie Hecht and Kevin Kay had stumbled onto the company's discarded MTV pilot—the one that inadvertently sent

Morgan Tilford over the precipice. At the time, the pair was remaking a new Viacom network, TNN, which had begun life as the country music–themed The Nashville Network in 1983. Hecht and Kay were converting it into The National Network, a pop culture station that would feature professional wrestling and *Star Trek* reruns.

Robot Wars, the pair realized, would perfectly complement the WWF programming. Instead of the inscrutable task-based competitions like Dean Kamen's FIRST, these were easily understood, smash-em, bash-em contests. They scooped up the option from their MTV counterparts and negotiated a license to film six episodes with the American builders and their robots in England. The U.S. presentation would be nearly identical to the U.K. version—a stylized mechanical game show, complete with the house robots like Sir Killalot, and the theatrical, fiery set.

Robot Wars would be late in entering the U.S. robotic combat market, but Hecht and Kay found a secret weapon. To host the new Robot Wars, they hired Mick Foley—the menacing, gap-toothed warrior primarily known as "Mankind" during his prolific 12-year career as a professional wrestler. Professional wrestling was the original "sports entertainment"—a staged athletic spectacle—and by picking Foley as the new host, Viacom, like the BBC in England, was putting Robot Wars in the same category. TNN executive Diane Robina put it bluntly in the network press release announcing Foley's involvement. "He's beloved by World Wrestling Federation fans, and is a natural to host this high-octane robot battle, featuring the house robots that develop as characters—much like the many characters of the WWF." Accordingly, the network planned to launch the show before its popular *Raw is War* wrestling program, leveraging the WWF fan base.

The biggest challenge facing Mentorn was getting American robot builders to participate. Most of the U.S. veterans still had deep allegiances to Battlebots and an aversion for anything associated with Steve Plotnicki. Still, there were new American builders who had been attracted to the sport by Battlebots, who didn't know and didn't care about all the bad blood. Mentorn realized it needed a U.S. liaison, someone inside the community who could coordinate with these unaligned builders and bring them into the Robot Wars fold.

Dan Danknick accepted the assignment.

It didn't bother him that Robot Wars was challenging Battlebots, that it still operated in conjunction with Profile, or that Steve Plotnicki—the guy he had disliked so much in the early years of the dispute—stood behind the scenes, working to guide the return of Robot Wars to the States. After his cancer bout, Danknick was taking a more dispassionate view. "It started out as a fun event," he said. "But I finally realized, it's just a business. We're not doing a public service or anything."

Up in Novato, Trey Roski heard about Danknick's new gig through the informal chain of gossiping robot builders. He fumed. Danknick had been a friend, one of the builders who sat in his living room after one of Marc Thorpe's bankruptcy hearings and urged him to start Battlebots.

Danknick's decision also challenged several of Roski's most fervently held assumptions. Roski figured that by bearing the legal costs to stand up to Plotnicki and restart the sport in the United States, he had acquired the robot community's undivided allegiance. By creating an event and a TV show that financially benefited the builders, he expected to maintain that loyalty. And after the U.S. Robot Wars event was canceled in '99, he thought he was finished with Mentorn and Profile forever—at least on U.S. television, if not in the courts. Danknick's new job, coupled with Viacom's announcement, opened Trey Roski's eyes. His comprehension of the challenges he now faced, stoked by his notorious temper, manifested itself as a disproportionate anger aimed directly at Dan Danknick.

Oblivious to all this, Danknick called the Battlebots office a few weeks before the May Treasure Island competition. He planned to bring thousands of dollars worth of fasteners, bolts, and electrical parts to the tournament, and wanted to make sure there would be room in the pits to set up a Team Delta table.

Roski lit into him. "I hear you're working for Robot Wars, Dan," he said. "I hear you're hanging out with Plotnicki in London."

In fact, Danknick had only exchanged e-mail with Plotnicki, and had never met him in person. But he admitted that he was serving as Mentorn's chief technical liaison in the United States. As it continued, the conversation grew heated. "There's no way in hell you can come to recruit for Robot Wars under the guise of selling a few trinkets," Roski said.

Danknick insisted he wasn't recruiting, just coordinating, and pointed out that like Roski, he was merely running a business, trying to further his fortunes. And he couldn't believe Roski didn't respect the Team Delta business, pointing out that his robot parts were purposely engineered to improve the safety of the sport.

But Roski thought Danknick's behavior constituted treachery. He banned Danknick from that spring's Battlebots event and slammed down the phone.

Danknick stayed away from Treasure Island, and began actively helping not just Robot Wars but all of Battlebots' new rivals. That spring, he helped The Learning Channel intensify its production of Nash Entertainment's *Robotica* by building a drone robot to test lighting and camera angles. When he heard The Learning Channel was looking for an on-air technical analyst for future seasons, he sent the videotape from the Las Vegas Battlebots pay-per-view event, which he had cohosted. He enjoyed the irony. "Trey paid for my demo tape. I'm sure he'll love that," he said.

At the Battlebots office in Novato, Roski stewed over Danknick's perfidy. During an interview that summer, he produced a potato-shaped canine chew-toy and waved the doll back and forth over one of his giant 85-pound weimaraner dogs. "Here boy," he cried out. "Get Dan! Get Dan!"

◘

"I got to tell ya, Ahmed."

Wrong host, start over.

"I got to tell ya, Tanya."

That's right.

"Wizard of Saws has free rein over . . ."

Wrong line.

"Ill Tempered Bot has a lot of speed but it won't help him here."

Uh, hold on. Speak into the mike, Dan.

"Sound check, one, two, three . . ."

Danknick was standing in front of the camera, microphone in

hand. Nash Entertainment had given him the job of on-air technical correspondent for Robotica, where he complemented the over-the-top antics of host Ahmed Zappa—son of Frank—and Tanya Memme. As the tech guy, Danknick explained the inner workings of various bots to the TLC audience. It was September 2001, and they were taping seven episodes each for seasons two and three at the Hollywood Center Studios—a fenced-in complex of sets and soundstages in the middle of Hollywood.

Danknick was working for Bruce Nash, a veteran in the burgeoning television field of "reality entertainment." Nash was the creator of such critically maligned fare as "Magic's Biggest Secrets Finally Revealed" and "When Good Pets Go Bad." Back in 2000, he had noticed the success of *Robot Wars* in the United Kingdom and *Battlebots* in the States, and had an employee write Tom Gutteridge to ask about licensing the *Robot Wars* spinoff *Techno Games.* When Gutteridge declined, Nash jumped at the chance to put his own stamp on the innovative TV genre.

To help with the show, Nash found a producer named Andrew Greenberger. Ironically, Greenberger had helped film a 1998 documentary on the FIRST competition and was totally captivated by Dean Kamen's robot contest. "It's a wonderful cause, almost like a river with tributaries," he said. "The impact of it will affect everybody in so many positive ways." But like the Nickelodeon guys, Greenberger ultimately concluded that FIRST was too "soft" for TV. He joined with Bruce Nash to create a robot competition more presentable to a mainstream audience.

Each episode of *Robotica* featured six robots attempting to negotiate two obstacle courses called the Gauntlet and the Labyrinth. All six teams were the subject of a short, news-style feature on their background and robot-building operations, and at the end of the episode, the best two performing bots of the group fought head to head, sumo-style, on a raised platform.

Robotica, Nash said that week from a comfortable break room at the studio, "Is more of a thinking person's show. It asks builders to take a machine and adapt them to a number of different challenges." Unlike *Battlebots, Robotica* did not allow everyone to participate. *Robotica* staff picked 24 teams per season to compete. The longtime veterans were mostly unavailable; they were loyal to Trey Roski and didn't want to help other shows. A few had experimented with *Robotica* in its first season earlier in the year, like L.A. builders Christian Carlberg and Jason Bartis, who had competed with Bartis's new robot, Mini Inferno. They said they felt bad afterward and didn't sign up to compete in *Robotica* again. "I felt an allegiance to Battlebots," Bartis said. "They proved themselves to us through a few years of hard work."

But even without sustained interest from the old-timers, Nash had few problems signing up enough qualified teams. In the wake of the sport's popularity, there was, as anticipated, now a new camp of builders who didn't know about the sport's partisan politics and were enthusiastic about pursuing any opportunity to compete. Many were Battlebots newcomers, rookies at the Treasure Island competition enticed by the unique challenges of *Robotica,* the TV exposure, and $10,000 in prizes.

Team Whyachi was in this new group: Terry Ewert, his wife Lisa, and Dale Hammel went to *Robotica* thinking it would help get the Whyachi team name out to the new robotic-sports-loving public. If the sport truly was evolving to NASCAR proportions, Ewert thought, perhaps *Robotica* was going to be the Bristol Motor Speedway to the Battlebots' Daytona 500. Why shouldn't he compete at both?

That week, trapped in the studio, Ewert regretted the decision. Unlike the fast pace of a Battlebots tournament, where one match followed another in quick succession, *Robotica* was tedious and dull. Sure, the Gauntlet and Labyrinth sets were cool, featuring breakable panes of glass and a faux-industrial look. Also, the catered food was excellent, and Nash's company put the builders up in nice hotels across town and paid for everything. But the waiting—the waiting was brutal. Hours passed between matches, as producers reset the lights and Dan Danknick and the other hosts did take after take in front of the camera. Hundreds of stagehands swarmed the set and kept each team under constant, annoying surveillance. Competitors were allowed access to their bots for only 45 minutes between bouts, better to preserve the sanctity of the pits, which itself was part of the set. "This is just like the old folks' home," one builder quipped. "We eat and watch TV."

Ewert in particular found the inactivity oppressive. "This isn't a sport," he spat between cigarettes out in the courtyard. "It's a TV show that just happens to have bots in it." Back home, his employees at Westar were working on three new robots for the next Battlebots tournament, and Ewert desperately wanted to return to oversee the work. Perhaps not coincidentally, his *Robotica* entry, Whyatica, fared poorly. Though well-built, with a titanium chassis and a V-shaped plow in front, it was too fast and offered too little control for Ewert to properly navigate the mazelike Gauntlet. In the second event, the Labyrinth, something triggered one of the failsafe mechanisms in Whyatica's speed controller and the robot didn't even move. Two other competitors moved on to face each other in the finale of that episode, and Ewert was free to return to Dorchester. He was pleased. "I really want to get the hell out of here," he said.

But Ewert's view on *Robotica* wasn't shared by everyone.

Mike Konshak, another *Robotica* contestant, was a gray-haired Vietnam veteran and former motorcycle racer. When he wasn't participating in mechanical sports, he designed disc-drive arrays for Louisville, Colorado–based Storage Technology, and he held 14 patents. Like Terry Ewert, he was part of a new generation bringing ample resources to the sport that didn't have the same exclusive loyalty to Battlebots.

In 2000, Konshak broke his foot racing motorcycles.On Christmas Day that year, he stumbled onto the TV show *Battlebots* and was hooked. Motorcycle racing was for the young, anyway. With robotic combat, he could satisfy his mechanical curiosity, compete in life-or-death situations, and put the robot—not himself—at risk.

For his first Battlebots event that May on Treasure Island Konshak built Pyramadroid, a variation on the slinking, skirted Biohazard. Konshak said he "sadly underestimated" the beating that robots took in the Battlebox, and Pyramadroid got blown out in the preliminaries.

The defeat was depressing. Konshak had spent all that money on his robot, travel expenses for himself and his team, and on the new, rather expensive entry fee for rookie robots. He also felt there was a "good-old-boys network" of veterans who got preferred status. "They get seeded right into the televised finals," he said. "It's not quite fair. Some guys have robots that are not quite as good as ours." Though he loved the camaraderie in the Battlebots pits, *Robotica,* he felt, was more egalitarian.

Konshak's *Robotica* robot, Flexy Flyer, sported four all-terrain tires and a lifting arm, and could be reconfigured in three different ways depending on which course it was attempting. Flexy Flyer handily won the second season of *Robotica,* defeating a four-wheeler with a lifting arm called Killa Gorilla in the finals.

In the courtyard outside the studio that week, Dan Danknick could barely conceal his satisfaction at the promise of the new show. The mood during the filming might not be vivacious, he conceded, but the show would be better than the jokey, innuendo-laced presentation of *Battlebots. Robotica* was made specifically for television: The robots were selected for variety and showmanship, the sets were entertaining, and the tone was smart. "I see this as a middle ground between *Battlebots* and FIRST," Danknick said. "It's about engineering, not beating each other senseless. I think this will squeeze *Battlebots* right out of existence."

ﾛ

The toys hit shelves that summer. There, in all its miniaturized plastic molded glory, was the deadly blade of Jim Smentkowski's Nightmare . . . attached to the stylized base of Donald Hutson's heavyweight Tazbot. There were the metallic gripping jaws of Hutson's super heavyweight Diesector . . . transposed onto the body of Mark Setrakian's Mechadon. The bladed wheels of Trey Roski's Ginsu—on a robot that the toy packaging called Berzerker Bot. The whirling dome of Ziggo, augmented with a metallic hammer—called, bizarrely, Torque-A-Motor.

What was going on?

The manufacturer, the Schaumburg, Illinois–based A-Ha Toys, wasn't listed in directory assistance and its attorney didn't return phone calls. It had clearly not heeded the Battlebots advertisement from earlier that year, warning off copycatters with the line, "If you think our robots are tough, wait until you see our lawyers." A-Ha's line of toy key chains, dubbed Robo Battles, flooded drugstore shelves nationwide at the same time the first wave of *real* Battlebots toys came out, manufactured by the Malibu-based Jakks Pacific. These were key chains and "grip n' grapplers" (pull the trigger, and a tiny cable moves the harmless weapon). Unlike the A-Ha toys, they were reflective of the real thing, miniature versions of robots like Toro, Mauler 51-50, and Grendel, and the builders pocketed a small percentage of each sale. Roski's lawyers at Latham and Watkins slapped A-Ha with a lawsuit in the central district of Los Angeles federal court, alleging copyright infringement and trademark dilution and asking for a preliminary injunction.

The attorneys for A-Ha Toys responded by questioning whether robot features like bladed wheels and gripping jaws were truly unique, and also suggested that Battlebots hadn't properly trademarked the specific design features of each robot.

Trey Roski scoffed at this. The appropriation of his Ginsu made him particularly mad. "Before Ginsu, no one had thought of saw blades as wheels," he said. "I mean, come on. These are obviously rip-offs."

Much was at stake: The future of his business relied on the successful sale of toys to reimburse the builders and subsidize the continued improvement of their robots, which in turn would keep viewers interested and enthused.

At first, federal judge Robert Timlin sided with A-Ha. "While there is some degree of similarity of ideas and expressions between the two companies' products, an ordinary, reasonable observer would not find a substantial similarity." Timlin also found that A-Ha would face "economic ruin" under a restraining order, while Battlebots would not be greatly harmed by A-Ha's products. A year later though, the parties settled the dispute and A-Ha agreed to stop shipping its toys.

Nevertheless, Roski was pissed. In a temporary fit, he blamed his $350-an-hour Latham and Watkins attorneys for not properly trademarking all the distinct robotic features, and wondered aloud if he had the best possible legal representation.

To top it all off, the toys from Jakks Pacific were selling only modestly well. Most builders thought the quality of the toys was poor and noted that, ironically, the A-Ha toys actually looked cooler. Caleb Chung, a participant at the first Robot Wars in 1994 and the inventor of the Furby doll, had another explanation. "The appeal of the brand is based on taking 100 pounds of metal and servos and ramming them at 40 miles per hour," he said. "It makes a different sound than 30 ounces of plastic smacking another piece of plastic."

There was also another troubling possibility: that a television show that aired on a school night, at ten o'clock, on a network that spiked each broadcast with advertisements for programs like the racy *The Man Show,* simply wasn't reaching the kind of young audience that would be buying toys in the first place.

Meanwhile, the more expensive—and more promising—remote-controlled Battlebots from the Tiger Electronics division of Hasbro were due in time for Christmas. But as early as that August, Roski began getting worrisome phone calls. "The head of nearly every major toy store called to tell me it looked like Hasbro would be late and miss the holidays," he said. "I kept reassuring them we would make it." It turned out the toy honchos were correct: Hasbro's Battlebots toys wouldn't hit stores until January 2002, well after the holiday buying season, and they would also meet with a lackluster customer response.

ꟼ

Meanwhile, Battlebots' legal entanglements were deepening. Back in 1999, Steve Plotnicki had sued the Roskis in New York federal court, accusing them of interfering with Robot Wars' U.S. buisiness. Plotnicki dropped the case before a skeptical Judge Jed Rakoff, after Rakoff refused to grant a preliminary injunction to stop the upstart competition.

Now Plotnicki was backIn the summer of 2001, the music exec renewed all of his claims in the same court, using ammunition he had gained through the examination of the Thorpe family records in Santa Rosa. "Our beef with Trey Roski is that Marc Thorpe was clearly on his payroll," Plotnicki said. "He was paying Marc Thorpe's legal fees. And the result of him spending that money was that the American portion of our business was stuck in the mud." Plotnicki, initiating the suit in the name of Robot Wars LLC, restated all of his allegations in two separate and entirely new cases, one against Roski—and one against Marc Thorpe.

The new cases were filed before Federal Judge Gerard E. Lynch, the fourth trial court judge to consider the mass of allegations and intrigue that stemmed from an annual event that happened just four times in San Francisco during the nineties.

As if Roski's life wasn't difficult enough in the cool summer of 2001, he also had to deal with Barrett Lyon, a Sacramento-based hacker who registered the Web address Battlebots.org. Roski's attorneys sent Lyon a cease-and-desist letter, asking him to turn over the domain name; Lyon argued he was using the site in the context of "Internet bots," bits of software code that inhabit the online chat network called IRC or Internet Relay Chat. Instead of surrendering the domain, Lyon appealed to members of the open-source Internet community, urging them to send e-mail to Battlebots accusing it of being "humorless" and "comedy impaired." For a few weeks in September before Lyon caved in and handed over the URL for $70, the Battlebots office in Novato was inundated with castigating e-mail, accusing them of aggressive, litigious behavior.

Though the matter was a relatively minor one, the Battlebots CEO was getting a crash course in the challenges of running a business. Most disappointing was that the new robot competition shows seemed to be diluting the overall popularity of the genre. The second-season episodes of TLC's *Robotica* and TNN's *Robot Wars* garnered an average of 409,000 and 339,000 households respectively, while the audience for the third-season shows of Comedy Central's *Battlebots* had declined to about 313,000—a significant dip from the early days, when nearly two million tuned in for each show. Overseas, *Battlebots* got a summer slot on the BBC2 in Great Britain, but the show was canceled after a month of poor ratings. The TV audience in England was simply more accustomed to Mentorn's "sports entertainment" version. Meanwhile, Mentorn's *Robot Wars* was airing in 27 different countries around the world.

So here was the view from the other side of the intellectual property fence: disloyal friends, knockoff toys, cyber squatters, erratic TV ratings, and unreliable toy companies. By the start of fall 2001, Battlebots was running $100,000 a month in legal bills. "Call me only if you have to," Trey Roski told his lawyers.

The second Battlebots event of 2001 was held that November, in another naval hangar on Treasure Island. There was another giant, enthusiastic crowd of gearheads and 350 teams, a slight decline from the previous May attributed to travel concerns after September 11. But there was just as much roiling, pent-up enthusiasm in the pits. Veteran builders like Carlo Bertocchini, Jim Smentkowski, and then–heavyweight champ Terry Ewert signed autographs for members of the crowd. Game designers walked the floors, evaluating the robots for new videogames for Playstation and Gameboy. Ring announcer Mark Beiro and *Battlebots* host Bil Dwyer walked through the crowd, shaking hands with spectators. Old friends and rivals shook hands in the pits and talked about their robot-building efforts and prospects for the sport.

Even though the toys weren't flying off shelves, a few teams enjoyed the thrill of their first royalty checks from sale of the Jakks Pacific products. Rumors of the dollar figures rippled throughout the pits: Most of the builders with toys got between $5,000 and $15,000. Donald Hutson received a check for more than $50,000. His stylized robots, Tazbot and Diesector, were proving particularly appealing to toy store customers.

Christian Carlberg got one of the more average checks. Demanding and ambitious, he wanted to know exactly how many units his robots had sold and wondered aloud if the 15 percent royalty was fair. "At some point, over time, they should bump us up to at least 25 percent," he said.

Meanwhile, Comedy Central's response to flagging ratings was to bring in a new botbabe: Carmen Electra. The former *Baywatch* star and Playboy model was mostly known for her volatile marriage to basketball star Dennis Rodman. She too wandered the pit area that week, wearing a series of revealing outfits, flanked by a cadre of makeup artists. But the builders had learned to ignore the distractions of the TV coverage. Unlike *Robotica* or *Robot Wars* in the United Kingdom, this was a real tournament, with only the overlay of a TV production. Matches followed each other in quick succession. Teams fought each day, sometimes twice a day, and there was always something going on, and new innovative robots to inspect.

Donald Hutson brought a new middleweight robot, called Root Canal, which featured an inventive steering system called omnidirectional control. It used four sets of barrel-shaped rubber beads, set at slight angles to the floor, to allow the robot to move up, across—and diagonally, "Just like you're playing Quake or Doom," he said. The robot lost early in the TV rounds, but for Hutson, the victory lay in breaking new ground. "I wanted to be the first to do a good omnibot that worked," he said. "I won the minute it came into the arena and all the other builders said, wow, that's cool." Hutson vowed he would return to the competition next time with a strengthened robot. "The more we get kicked down, the more we try to perfect our robots. That's the difference between us and other people. That's what being mechanically inclined is really all about."

It wasn't as easy as it sounded. The previous May, Stephen Felk's Voltronic (he changed the name from "Voltarc" when he discovered that the company that made the neon bulbs for elevators owned the URL) lost listlessly to Christian Carlberg's Overkill in the quarterfinals. Between competitions, Felk added new foam-filled tires and new miniature front-mounted wheels called casters, to try to improve his bot's mobility. But the heavyweight had taken too much damage over its four-year life. In its first fight that November, Voltronic faced a red, four-wheel bot called MOE, short for Marvel of Engineering. It featured a thick and dangerous protruding metal propeller and was created by Cedar Falls, Iowa, engineer Norm Muzzy, who worked for John Deere and claimed expertise in pyrotechnics.

Felk planned to attack quickly, to get around the propeller and lift MOE into the air, just as he had with Biohazard two years before. Instead, Voltronic moved sluggishly, "As if it was wading through molasses," Felk said. MOE attacked and won, 36 to 9. Felk said he was actually relieved. "The robot gods are saying, it's time for a new robot. I'm a big believer in the robot gods. When they talk to you, you listen."

As Fuzzy Mauldin said that weekend, "You have to innovate to stay in the game."

That was Terry Ewert's approach. After the last competition, Battlebots had changed the rules to decrease the walker weight advantage that Ewert had so cleverly exploited. That meant the super heavyweight Whyachi was too heavy to compete, and Son of Whyachi became the team's super heavyweight entry. Ewert responded to the change by building a new heavyweight and middleweight, both spinners, identical in design. Both were hexagon-shaped, made entirely of titanium, and carried their components—and their weight—on the periphery. When they spun, using an independently controlled foot at their center for directional reference, these new spinners threatened to deliver a mighty wallop of kinetic energy, the unholy marriage of mass and speed.

Unfortunately, it didn't work out quite as well as Ewert hoped. For each shot they dished out, heavyweight Y-Pout and middleweight WhyNot took an equal and opposite blow to their own frames. And since all of the sensitive components were on the edge, they were especially vulnerable. Both robots lost their first fight: Y-Pout got up to a full-speed spin of 900 rotations per minute, and seconds after the initial cataclysmic collision, came to a dead standstill. The battery pack kept on going and ejected from the robot. Later, the middleweight fought a simple, boxy contraption called MaddGoth. After its first hit, WhyNot ricocheted like a hockey puck into the Lexan wall and expired.

After each match, the crowd went wild, stomping and cheering. Even after Son of Whyachi's amazing win over Biohazard at the last competition, the builders were back to applauding the Wisconsin team's bad fortune. "Why are you the only team in Battlebots history that everyone loves to hate?" another builder asked Terry Ewert after a match.

They were "hated" because, for all their good-natured Wisconsin enthusiasm, they still represented to many builders a wealthy team that might elevate the sport out of the hands of the garage builders. "It's the dawn of the big age teams," said 30-year-old Oakland video game programmer Paul Mathus, the builder of middleweight Das Bot, in between naps in the pits. Mathus worked alone and was recovering from several sleep-deprived weeks retrofitting his middleweight, which lost early in the competition. "How can I compete with the big teams? I'm one guy," he said.

Many of the new teams looked a lot like Team Whyachi: large groups of professionals who watched the show every week and were energized by the engineering challenge. For every obsessed hobbyist like Stephen Felk and Paul Mathus, who toiled alone over their ghettobots, there was a Team Jabberwock, a crew of six nuclear engineers from the Savannah River Power Plant in Aiken, South Carolina. Team leader Clyde Ward said he got 20 grand to build the robot from Westinghouse, the plant's parent company, which operated under a contract from the Department of Energy. Before each match, the nuclear engineers called their colleagues back at the plant. Over the phone, they talked about how to modify the trapezoidal, spike-armed Jabberwock to face specific opponents, like the spinning Mauler 51-50.

In that fight, Jabberwock didn't need much modification. Morgan drove, while the Supreme Commander stood nearby, dressed in a full-size chicken costume. Mauler 51-50 took one shot and flopped over onto its back and spent the rest of the match spinning futilely. Afterward, a subdued Morgan Tilford didn't seem to mind much. "Losing is part of the competition," he told the TV cameras. At his pit table, he was keeping a collection of planted orchids and ferns. He planned to give them to Trey Roski, as a gesture of friendship.

Jabberwock sailed into the semifinals, where it faced mighty Biohazard.

Since his utter destruction at the hands of Son of Whyachi last season, Carlo Bertocchini had completely rebuilt the former champ. From the outside, it looked like the same low-riding, titanium-skirted lifter, but Bertocchini had spent two months replacing bulkheads, installing higher-capacity speed controllers for each set of wheels instead of only one, and replacing the armor that Son of Whyachi had torn off. Everything electrical was new, and capable of withstanding even more damage.

To face Biohazard, the nuclear engineers added a pneumatic ram with a hard steel point, and sensors so it would fire automatically when the robots engaged. The device worked well, shattering the hinged blade on the tip of Biohazard's lifting arm, but Biohazard dominated the rest of the fight, flipping Jabberwock repeatedly. In the pits after the match, Clyde Ward vowed revenge: "We'll be back and well be better. Our goal is to never make the same mistake twice."

Bertocchini moved on to face Nightmare in the quarterfinals. Smentkowski's massive blade dug into Biohazard's titanium armor twice, sending up a pretty but ineffectual shower of sparks, then stopped spinning altogether. Bertocchini had his way for the rest of the match. In the semifinals, he faced Tazbot, and Donald Hutson followed a long line of competitors who had tried and failed to out-leverage the sneaky low-rider. At one point, Tazbot got marooned on its flat base, whereupon Bertocchini carted Tazbot off to the Pulverizer. Biohazard won 24 to 21. Bertocchini wore a delighted grin back to the pits, to prepare to recapture his crown in the finals.

The championship match was fought against Christian Carlberg's Overkill. Carlberg waged the fight brilliantly: Instead of trying, and failing, to get under Biohazard, he simply pounded Overkill's blade back and forth against the floor. Every time the blade hit Biohazard in the process, Carlberg scored points. Halfway through the match, it looked as if the strategy might work. Biohazard's arm got stuck halfway down, and Overkill was being more aggressive, but Bertocchini didn't look worried, and in fact, he had no reason to be. The stuck arm allowed him to push Overkill around the ring, delivering it to the kill saws and pushing it up against the arena walls. The fight went the whole three minutes. Afterward, in front of an exhausted but still raucous crowd, Mark Beiro announced that Biohazard had recaptured its crown, 27–18.

With Tony Buchignani, whose Hazard won the middleweight category, and Jonathan Ridder's Ziggo, which collected the lightweight nut for the second time, Bertocchini had demonstrated that the age of the lone gearhead was not yet over. At least for now, the ghettobots had the same chance as the big teams. The power of idea, the ability of the builder, and the element of fortune still ruled the robotic roost.

In the end, a group of American robot builders returned to Robot Wars, but it wasn't in San Francisco and little remained the same. Fifty U.S. teams got on airplanes, with tickets paid for by Mentorn, and flew to London's Heathrow Airport. Mentorn put them up at an airport hotel and bused them to the Shepperton Studios, a gated complex of TV sound stages 45 miles outside the rain-drenched bog of London in January. Tom Gutteridge's people also arranged to ship all the robots as freight. There wasn't any prize money and Mentorn didn't let the builders keep the trophies, but for their time and effort, each team received $2,000 in cash.

Like the producers of *Robotica,* Mentorn hand-picked the robots that would participate from applications it received over its website. Terry Ewert of Team Whyachi applied with the former champ Son of Whyachi; Ewert heard Mentorn needed good robots and wanted to exhibit his walking helicopter spinner one last time, before he replaced the walking mechanism with wheels. But the bot that had pulverized the champion Biohazard nine months before received a curt rejection over e-mail. Mentorn likely didn't want to use a robot that was so prominently associated with its rival Battlebots, but Ewert said, "I suspect they didn't want to repair the house robots."

Most of the builders that Mentorn chose were relative newcomers, with a year or less of Battlebots experience. A sword swallower and circus performer from Las Vegas named Ed Robinson brought a swordfish-shaped robot called Snookums—basically a scavenged propane tank on wheels, which housed some haphazardly assembled electronics. Robinson's bots perished quickly in the last two Battlebots tournaments and in the taping of the second season of *Robotica.* But Mentorn must have liked Robinson's gaudy pirate costumes, and Robinson leaped at the opportunity to perform. "I have more chance to stand in front of a camera here and act like an idiot," he explained. "If you don't like wearing silly costumes and acting like a fool, you might as well not come to Robot Wars."

Complementing the newcomers was a smaller group of veterans, including Todd Mendenhall, Bob Pitzer, and Patrick Campbell. Each had different reasons for wandering into territory where many of their fellow builders refused to tread. Some wanted to make sure they were involved with whichever aspect of the phenomenon ended up winning out; others weren't benefiting from Trey Roski's toy royalties or getting enough exposure on TV to make the effort worthwhile. L.A.-based special-effects worker Val Sykes summarized his philosophy this way: "I have no allegiances. I have an allegiance to me. It's hard to be a first-timer at Battlebots. The ranking system, the preliminaries—it takes away the fun real fast."

Bob Pitzer was harsher. "Trey is trying to prove something to his dad, but the writing is on the wall now. Robot Wars will come into the United States with Viacom's backing, and Comedy Central won't matter." With his brother Chuck, Pitzer was behind Team Raptor and the annual Botbash tournament in Phoenix, Arizona, a smaller competition that wasn't televised. A robotic combat entrepreneur in the same vein as Dan Danknick, Pitzer's loyalty to Roski was beginning to erode. "I'm making friends here. I'm trying to bring Team Raptor into every aspect of this sport," he said. "I don't owe anybody anything. I've spent tons, and nobody is really paying us back completely yet."

Danknick was there too, in the "coordinator" role that had spawned his simmering feud with Battlebots. High on the prospects of *Robotica* only a few months before, Danknick now figured the future of robotic combat rested with the theatrical interpretation of Mentorn's *Robot Wars.* He speculated Battlebots didn't have much longer to live and said he wouldn't lose much sleep over it. "I'll regret that Greg Munson will lose his job, because Trey didn't get with the program," he said. "*Battlebots* may be very good competition, but it's a sucky TV show."

Back in New York, Steve Plotnicki enjoyed the irony of a Robot Wars TV show with American competitors. "These are the robot builders Trey Roski thought he controlled," he said.

Mentorn, it seemed, had learned a lesson from the British roboteers' complaints, and the American builders were treated well. The U.S. competitors ate three free meals a day, in double-decker buses furnished with booths and tables. Tech crews and production staff swarmed the pit area, helping teams get ready for their matches. To conform to British law, the Mentorn tech crews outfitted each bot with a specialized 40-Mhz radio and a safety switch.

The show was going to be nearly identical to the U.K. version:

Each episode would feature a slightly different version of robot combat, including tag-team events, melees, and one-on-one face-offs. As in the British show, the house robots like Shunt and Sir Killalot stood ready to inflict massive damage on losing contestants, but when a competitor lost its match and got demolished by the house robots, the tech crew helped the builders put their robot back together, so it could go right back into the arena and fight again in another episode.

Mentorn taped all these different competitions to allow Viacom to package the footage into a series of 13-hour episodes for TNN. This time, they also filmed six half-hour shows for sister network Nickelodeon.

Little about the taping indicated an authentic competition. It was a TV show, pure "sports entertainment," with the builders and their robots filling the roles of actors in a mechanical melodrama. Before matches, the competitors were lined up in front of a tunnel leading to the arena and instructed to playfully yell at each other—professional wrestling style.

Live audiences of 900 8- to 10-year-old kids and their dads tramped into Shepperton Studios, three times a day, to watch the show. The kids wore *Robot Wars* T-shirts and *Robot Wars* sneakers, carried *Robot Wars* tote bags and foam fingers that said "Pit" on them, a reference to the arena hole where most contestant robots ended at the end of each match. A few expressed confusion, thinking they were coming to watch the familiar British contestants, but they were quickly mollified by the free souvenir programs handed out to each of them, featuring full-color photos of the famous house robots and an interview with their creator, BBC special-effects chief Chris Reynolds. During the half-hour delays between fights, the audience was treated to videos of the British show, plus appearances by the house robots and a warm-up host quizzing them with *Robot Wars* trivia. "What is Sir Killalot's main weapon?" And, "Mick Foley, who is a presenter on *Robot Wars,* is also famous for what?" The kids knew all the answers.

After each session, they flocked into the studio next door and lined up in front of a merchandise stand and gathered around 14 blue Sony Playstation kiosks, where they played the new *Robot Wars* video game.

The American builders suddenly found themselves signing autographs and placing spare pieces of robot armor into outstretched hands. The furor far surpassed *Battlebots* mania in the States. "It's a different feeling over here," said Todd Mendenhall. "Robotic combat is an established institution."

The sport had come so very far since 17 prototype robots competed in San Francisco nearly a decade before.

That month, the British newspaper *The Guardian* ran an article proclaiming, "Mentorn's robots dominate the globe." The paper quoted Tom Gutteridge about his worldwide success and reported that the show had generated more than 100 million pounds in sales and merchandising revenue around the world. Mentorn had recently sold the show to China and Australia and was negotiating with Japanese networks. A toy deal in the United States was being negotiated, and an animated cartoon series was in the works that would feature the house robots.

Left unsaid was the fact that the creator of the sport, and most of its British participants, had yet to see a single pound.

Between filming sessions, Mick Foley wandered through the cold London drizzle back to his trailer, where he sat on a bench and read a novel.

Foley, 37, was a contradiction in terms. The former professional wrestler certainly looked like a brute—six feet two inches tall, 280 pounds, with one eviscerated ear and four missing teeth, all wounds from his career in counterfeit combat. But he was also the author of two best-selling autobiographical books and a forthcoming novel, and had the reputation of a gentle sage to all who worked with him. On the air, he rasped hoarsely about robot destruction ("Tonight, chaos reigns supreme!"). In the privacy of his trailer, he spoke in soft, thoughtful tones about the emerging genre of robotic competition. "*Robot Wars* occupies a place all its own," Foley said. "I don't know what to compare it to. Remote-control cars have been around a long time. Instead of racing them, you get to see which has more firepower."

Foley admitted that he was not himself mechanically inclined. The height of his abilities in that regard involved learning in an eleventh-grade auto-shop class how to jump-start a car, a skill he had long since forgotten. But he was learning a lot about robot-driving strategies, he said, and he enjoyed ribbing the introspective, science-minded contestants on the show.

Well, what did he think of that turbocharged audience of kids inside the arena next door? And how much credit for the whole thing did he give Marc Thorpe? Foley flashed his personable gap-toothed grin. "I liken that audience to a wrestling audience. Families get together to watch things be destroyed. Instead of wrestler's bodies, it's robots."

Perhaps not surprisingly, he didn't know who Marc Thorpe was.

Epilogue

In January 2002, the artists and anarchists that orbited the San Francisco robotic arts group SRL gathered to raise money for one of their injured comrades, Amy Miller.

Miller was the orange-haired SRL member who fell 20 feet off a trucking container during Mark Pauline's Berkeley show the year before. She broke her pelvic bone in three places, cracked her hip socket, and wouldn't be able to walk normally for months, let alone work and pay her rent. Even worse, Pauline's insurance carrier was resisting paying her huge medical bills. The SRL chief grumbled that he might have to sue the company.

At the benefit held in the South of Market Arts Gallery in San Francisco, a 30,000 square foot city-owned cultural center and exhibition space, attendees paid between $10 to $25 at the door to interact with an assortment of mechanical attractions and curiosities. Inside, guests could manipulate a seven-segment mechanical snake called The Grub, or a tethered model called the Tetrahedron, which changed shapes and flopped around a table by extending and retracting pneumatic tubes. In a fenced yard behind the gallery, longtime SRL collaborators The Seemen and PeopleHater invited guests to sit inside their various mechanical contraptions and get spun around or squirted with strange liquids.

Mark Pauline had wanted to install the Pitching Machine, which gunned planks of wood, alongside the contraptions of his fellow artists, but the city wouldn't let him.

During the party, Amy Miller wheeled around in a custom-made lounge-chair, greeting old friends and thanking well-wishers. There had been false rumors about the severity of her injury, so (playing along) she had a friend rig a fake plaster cast over her lower body, with gruesome metal traction rods sticking out of it. Actually, Miller was already ambling around on crutches.

"This is an outpouring of friendship and love that I always knew was here, but that rarely gets expressed," she said of the party in her honor, while handing out chocolates to the crowd. "Everyone is so busy with their smaller subgroups, it's very rare that we come together like this. It's a beautiful expression of how I see our community."

The main attraction of the evening was the premiere of the video of SRL's Berkeley show the year before. On a big screen before a rapt crowd, the whole apocalyptic scene repeated itself. The Running Machine once more nudged the Sneaky Soldier into the flaming bath of calcium hydride. The Track Robot shook its pincers at the bewildered crowd, and the Pulsejet rocked the walls of East Bay residents miles away with its 140 decibel blasts of noise. Then the Pitching Machine excoriated Toys 'R Us mascot Geoffrey the Giraffe, and the Flamethrower submersed it in flame. Somewhere off-camera, a police dispatcher was directing a fleet of squad cars to the scene.

Standing at the back of the crowd, Mark Pauline watched the video with satisfaction. The performance was another feather in his cap after 24 years of illegal, underground robot performances. He was so pleased that even the city's refusal to let him run the Pitching Machine that evening didn't seem to bother him. Pauline would just bide his time until the authorities let their guard down, and then SRL would strike again.

He would soon turn 50. But he said, "I think I act and feel a lot younger than most people. At least I have the benefit of a solid lifetime of fun and games behind me."

The next month, members of another gearhead faction came together to celebrate the end of Marc Thorpe's legal problems. They gathered in the Orinda, California home of Will Wright, longtime competitor and designer of the popular video game "The Sims," and they marveled at Thorpe's difficult journey over the past half decade.

Two years ago, it seemed like Thorpe's fight with his old partner

Steve Plotnicki would last forever. Judge Alan Jaroslovsky of Santa Rosa federal bankruptcy court found no conspiracy to harm Robot Wars in the U.S., but he ruled that Thorpe had breached his agreement with his lukewarm online postings to the robot builder Web forums. The judge reduced the cash payment Thorpe was due to receive from Plotnicki, and both sides appealed the ruling.

Thorpe's Santa Rosa lawyer, Dave Chandler, was seriously outnumbered, as Steve Plotnicki had at least three lawyers representing his interests: Fran Jacobs, a partner in the New York offices of national firm Duane, Morris & Heckscher, who had represented Plotnicki before; William Pascoe, a Marin, California-based bankruptcy attorney, and the point-person in Santa Rosa for Plotnicki; and Harvey Schochet, one of the most prominent bankruptcy attorneys in San Francisco, who was hired to consult on the increasingly complex case.

Plotnicki's team appealed Jaroslovsky's ruling to the U.S. District Court, and then the Ninth Circuit Court of Appeals. Chandler also appealed to the U.S. District Court. Back in Santa Rosa bankruptcy court, he finally confirmed a debtor's plan—a schedule for Thorpe to repay all his loans.

That's when Plotnicki attempted a jurisdictional end-run around the unsympathetic Jaroslovsky. In April 2001, he filed a new case against Trey Roski and Battlebots, on his own turf, in the federal court for the Southern District of New York. At the same time, Plotnicki filed a companion suit in New York against Marc Thorpe. In both cases, he was suing under the guise of Robot Wars LLC instead of his new company, Astor Holdings, seeking to recover all the money he had lost because *Battlebots,* not *Robot Wars,* had reached American television first. Both new suits alleged the same conspiracy between Thorpe, venture capitalist Bob Leppo, and Trey Roski.

In his workshop in Fairfax on April 22, 2001, Marc Thorpe was served the summons. He felt like a man dangling at the end of a fraying rope, getting kicked again and again from above.

But Dave Chandler saw an opening: He turned around and sued Plotnicki within the framework of the bankruptcy case, arguing that the New York lawsuit had violated a tenet of bankruptcy law called the "discharge injunction," which said that once a debtor's plan to repay his debts is confirmed, creditors are blocked from further litigation against the estate.

Chandler's new lawsuit named Plotnicki and his lawyer, Fran Jacobs, as defendants. "The Thorpes have been extremely stressed over the refusal of the defendants to recognize and abide by the discharge injunction," Chandler stated in his complaint. "Marc Thorpe has lost sleep. He has lost time from his work as a consultant." Chandler asked for hundreds of thousands of dollars in damages, in addition to punitive damages. He argued that Thorpe should be compensated for all the new concepts he was prevented from creating because of the distraction of protracted litigation. Considering how successful one Thorpe idea had become—robotic combat—Chandler theorized that the punitive damages could stretch into the millions.

The argument was preposterous, of course, but Plotnicki couldn't underestimate it. Jaroslovsky had tired of his legal maneuverings long ago, and was proving unpredictable. "Mr. Pascoe," the judge had told Plotnicki's lawyer in one characteristic moment, "I understand the history of this case is one of gamesmanship, but your client and the other entities related to Mr. Plotnicki have gone beyond the pale."

Dave Chandler pressed his advantage and used Plotnicki's own hardball tactics against him. He demanded to see all the e-mail, letters, and phone call records between Plotnicki and his lawyers. He wanted to know what they had known about the discharge injunction, and when they had known it. Did they deliberately violate the law in their rush to persecute Thorpe? Dave Chandler flew to New York and took the depositions of Plotnicki and attorney Fran Jacobs. The interviews lasted all day, and were openly hostile.

Chandler had targeted Plotnicki's most potent weapon: his lawyers. Perhaps not surprisingly, by the end of 2001, Plotnicki was ready to end his long battle with Thorpe. He said he saw no reason to continue, since the real damages would come in the case against the Roski family, if and when that suit ever reached trial. Explaining his decision to settle with Thorpe, Plotnicki said that Chandler and Thorpe "were still trying to fight like it was day one. We were able to forecast a couple moves down the chess board, where we might be able to yield."

Dave Chandler responded, "Of course we were fighting like it was day one. Marc's life was at stake."

With the extra leverage of the lawsuit, Chandler negotiated a new settlement that was even better than the old one. The artist got a single payment of $350,000 and ten percent of all Robot Wars LLC income. Of course, most of that money would vanish in an instant, paid back to friends, lawyers, and debtors. And even though Tom Gutteridge in London was telling newspapers about the many millions of pounds *Robot Wars* was earning around the world, Thorpe's cut would be surprisingly small. The pie, after all, was divided between Mentorn, the BBC, Plotnicki, merchandisers like toymakers, and last and least, the roboteers themselves.

Nevertheless, Thorpe's legal nightmare was over. He could move on.

For the past year, he had worked with Will Wright and Mike Winter in an East Bay warehouse, which they called "The Stupid Fun Club," filming a TV pilot in the go-motion animation style of the sixties cult TV program *Thunderbirds*. It featured wooden puppets who lived in a sixteenth-century Japanese village that had been altered by alien technology. A tank-like robot, created by Thorpe, roamed the village and caused trouble. The show was going to be named M.Y. Robot—they didn't know what the M.Y. stood for—but they were already represented by talent agents at the Creative Arts Agency in L.A.

When Wright and Winter learned that their friend's legal troubles had ended, they decided to celebrate the occasion in style. Wright sent out an e-mail invitation to the robotic combat community in the format of a news story, and included a 1945 photograph of Americans celebrating the end of World War II in Times Square. "End of Lawsuit!" the headline in the e-mail shouted. "Court Accepts Joint Settlement! VP Day Celebrated in Orinda!"

The party took place on a rainy night in February 2002 at Wright's home. Bob Leppo was there, chatting with Richard Friesen, the consultant who had originally introduced the venture capitalist to Thorpe. Trey Roski and Greg Munson were there, talking to friends and congratulating the artist. Now that Thorpe's case was over and the judge hadn't found any evidence of a conspiracy, Roski said he didn't feel particularly threatened by Plotnicki's lawsuit against Battlebots in New York. But since neither party seemed likely to surrender, that case too would probably go to trial at some point.

Veteran robot builders like Carlo Bertocchini, Stephen Felk, Jonathan Ridder, and Jim Smentkowski were also there. Christian Carlberg, Peter Abrahamson, and Jason Bartis drove up from Los Angeles.

Thorpe's wife and daughter were not there.

His lawyer, Dave Chandler, was the guest of honor. Chandler brought his wife and two teen-aged sons.

The party-goers mingled among the robotic curiosities in Wright's house. Roaming the living room was the $900 RoboScout personal robot from Sharper Image, plus a crawling cadre of Biobugs, intelligent toys designed to learn their environment as they scuttled around a room. On the shelves in the den, Wright had a large collection of Japanese anime videos. Lying next to them was a McMaster Carr catalog, the bible of the robot builder, and videos from the old Robot Wars events played on television. The builders clustered around and reminisced about each event fondly.

Later, the guests gathered around the den to hear from Thorpe himself. The 53-year-old Robot Wars founder seemed to shrink away from the heavy track lighting. "I always felt comfortable speaking at Robot Wars, but everyone was there for the event, not for me," he said.

"I feel so pleased that Will and Mike brought everybody together here to celebrate this," Thorpe told his friends. "I'm glad to be taking the road that leads away from litigation. I really don't recommend it to anybody." Everyone laughed. Thorpe then presented Chandler with a 14-pound trophy: a carbide tube topped by a brass cube, with four letters milled onto the four sides: HERO. A plate on the cream-colored nylon base of the tube read: "David N. Chandler, Attorney at Law, Case: Marc Thorpe vs. Robot Wars." Chandler was visibly moved.

Before Thorpe could resume his speech, Mike Winter started bringing in gifts. An old Robot Wars competitor named Jim Sellers had sent an unusual still-life sculpture: The centerpiece was a broken clear plastic cylinder and strategically placed around it were the keychain Battlebots toys of Ronin and Biohazard, making it appear that the robots were busting out of jail. "Hey Marc, congratulations, you're free at last," the card read.

"I'm overwhelmed," Thorpe said with tears in his eyes.

Mike Winter, his wife Becky, and their daughter Lisa presented the final gift: a homemade electric fortune-telling machine. Pushing on a plastic rod at the center of the device made a metal dial spin around and point to one of six small clay symbols which prophesized the future. The dollar sign, Winter explained, represented money; the heart represented love; the pair of dice signified luck; the scale was justice; the evil red face symbolized the devil. "We all know who the devil is, right?" Winter quipped to laughter.

Thorpe graciously took the gift and moved to press the plastic rod while the contraption was still in his arms. "No! No! Be careful!" Winter shouted. "It'll take your fingers off! It's basically a cuisinart without the guard!"

Thorpe put the fortune telling machine down, took a safe step back, and tried it. The guests all gathered around and peered in. The plastic dial spun around, past money and dice and love—and landed on the devil. "Oooh," everyone intoned.

There was some nervous laughter. But then the hosts brought out the double-layer, cinnamon-dusted, tiramisu cake, decorated with plastic robots, and someone started singing "Happy Birthday" with some creatively altered lyrics to reflect the situation, and Thorpe just stood there in the middle of it all, among the community that owed its very birth to him, and after a moment he was smiling again.

That same month, the first advertisement of Super Bowl XXXVI was a spoof on robotic combat: A heavily-armed champion robot was bludgeoned by a mini-fridge carrying a can of Bud Light, in an arena that combined features of Comedy Central's *Battlebots* and Mentorn's *Robot Wars*.

Bob Pitzer and his brother Chuck, longtime competitors, helped Anheuser-Busch's advertising agency craft the "Robobash" spot. The agency had approached the brothers the previous summer looking for an arena to use for filming, and the Pitzers turned the query into a week-long gig and a $30,000 fee to build and operate the killer robot that would take punishment from the Bud-fridge.

The Pitzers ended up using the base and hydraulic jaws of their super heavyweight Battlebot, TripultaRaptor, and added new arms and a Viking helmet. Fatefully, Bob Pitzer didn't mention anything about the commercial to Trey Roski. "We don't want Trey to have any part of it," he told friends. "He would just try to take control."

An estimated 84 million people watched the ad, and most media outlets ranked it among the best of the game.

Three months later, Battlebots sued Anheuser-Busch in the U.S. District Court in Los Angeles.

The complaint alleged that the player's agreement that the Pitzers had signed gave Battlebots the exclusive rights to commercially exploit the likeness of TripultaRaptor for five years; Anheuser-Busch, its advertising firm, knew this but decided to ignore it, and intentionally copied the Battlebots arena and style of competition. In addition to punitive and statutory damages, the suit asked for all the profits from Bud Light beer since the commercial started running.

Trey Roski conceded some of the damage demands were extreme, but said the suit was a necessary part of the entertainment business: "If you don't protect your trademark, then people take advantage of you," he said. He also pointed out that he had instructed his lawyers to specifically leave the Pitzers out of the lawsuit—he didn't want to walk down the same path as Steve Plotnicki, alienating the builder community by engaging its members in litigation.

It didn't matter: Pitzer was furious anyway. At the fifth Battlebots event in May 2002, held again on Treasure Island in San Francisco Bay, Pitzer walked the pits fulminating against the Battlebots CEO to anyone who would listen. "The whole point of us doing the commercial was to put the word out about robotic combat," he said. "The commercial was a market test. Budweiser was interested in getting involved. Then Roski sues them, the biggest advertiser in the United States. They were ready to start doing things that could have brought money into the sport and he shot himself in the foot."

Roski too was annoyed at Pitzer, whom he saw as a friend who had betrayed him. "He's made more money from us than he ever got or will get from Budweiser," Roski said a few days after the tournament. "And here he stabs us in the back. It's biting the hand that feeds you.

"The builders say to me, 'why does Donald make so much money?'" Roski continued, referring to the Tazbot and Diesector creator whose decorative robots were doing well on toy store shelves. "It's because Donald is *with* Battlebots. He's loyal and I'm loyal to him. The others go over to *Robotica* or *Robot Wars*."

Roski was beginning to sound like a stressed-out businessman who demands blind fealty from the legions around him. In a sport filled with fiercely independent, entrepreneurial gearheads, the approach was destined to generate conflict.

There was another reason for Roski to be in a bad mood. The Treasure Island competition, the fifth filmed by Comedy Central, had ended in disaster on the final day.

During the heavyweight rumble, a tooth on the upwardly spinning disc of Jim Smentkowski's Nightmare had come detached during the fight, and shot through an unprotected corner of the Battlebox ceiling. In a realization of all the greatest fears of competition organizers over the years, the shard of metal went spinning into the stands, where it scattered spectators and struck the shoulder of Battlebots organizer Greg Munson's sister—who happened to be holding her baby daughter, Munson's niece. Both were shaken but okay, but a horrified Roski and Munson canceled the remaining rumbles.

Then, in the final championship match between super heavyweights Vladiator and Diesector—which Comedy Central would use as the finale for the entire season—Vladiator careened so hard into a section of the Lexan wall that it popped the inch-thick "unbreakable" plastic from its track. Roski stopped the match and, walkie-talkie in hand, went to inspect the damage. Even though the Lexan popped right back in, he was too rattled to continue: He called the match with more than a minute left and ended the tournament without a clear winner. Lenny Stucker and the Comedy Central crew looked completely forlorn. *Battlebots* would soon be up for renewal. What might be the show's last season on the network had ended with a wimper.

"Are robotic sports here to stay?"

For the engineers and mechanical obsessives who populate the pits at the numerous robotic events around the world, the answer is an indisputable "yes."

Battlebots, Robot Wars, Robotica—the old TV shows will go away, but the interest they helped unleash in mechanical competition guarantees that new ones will pop up. For example, Mentorn's *Robot Wars* spin-off, *TechnoGames,* which features R/C robots in less-combative pursuits like rope-climbing and swimming, was recently picked up in the United States by the TechTV cable network. And The Learning Channel's *Junkyard Wars* and *Full Metal Challenge,* and The Discovery Channel's *Monster Garage,* continue to be unexpected hits. In these shows, gearheads compete to turn cars or other contraptions into extraordinary task-performing machines.

At universities around the world, design contests like MIT's 2.70 continue to thrive. Events like RoboCup, where the autonomous robots of various universities play soccer, and the firefighting contest at Trinity College, where autonomous robots compete to locate and extinguish a candle in a mock home, are helping to advance research into artificial intelligence. Those aren't spectator sports yet, but the Japanese organizers of RoboCup have vowed that by 2050, robot soccer players will be able to defeat the best human team.

Outside academia, the garage hobbyists drawn in by *Robot Wars* and *Battlebots* also keep robotic sports alive, gathering informally to compete in dozens of non-televised competitions. On the East Coast, an organization called NERC, "Northeast Robot Conflict," holds events for 12- and 30-pound robots four times a year. Its counterpart in the south, SECR, puts on similar events in South Carolina and Georgia. Bob Pitzer's Botbash, Steve Brown's Steel Conflict, and Ken Gentry's Robocide are also informal events that attract veterans of the sport. In the fall of 2002 in Pittsburg, Pennsylvania, Lycos creator and bot enthusiast Fuzzy Mauldin opened up the Robot Club and Grille, where you can order a hambuger—and fight with a rented combat robo. "It's the best meal you've ever had behind bullet-proof glass," the restaurant boasts.

In Great Britain, events like The Debenham Robot Rumble, and Sussex Robot Rumble are gathering spots for the competitors of Mentorn's televised spectacle. "Even if *Robot Wars* vanishes from our TV screens, the sport will no doubt live on," says Ian Watts, a veteran U.K. roboteer.

High-school robot competitions provide more fuel for the sport's development. In 2002, Dean Kamen's FIRST contest continued its impressive grass-roots growth. More than 650 teams competed in 18 regional events and in the finals outside EPCOT Center at Disney World. Kamen reverted to the old format of two robots competing directly against two robots—taking a step back from the idea of pure, nonviolent collaboration. "We listened to the students," he said, zipping around the Silicon Valley regionals on his newest invention: the Segway Human Transporter, a dual-wheeled electric transport which cost $100 million to develop and uses a complex system of gyroscopes, computer chips and software to mimic the way the human body maintains its balance.

That year, in what is surely a continuing disappointment for the famous inventor, the FIRST competition was broadcast on the seldomly watched NASA TV—the space agency's satellite feed.

The same weekend as the FIRST 2002 Silicon Valley regional, two other student-oriented robot competitions were taking place in different parts of the country.

In Texas, Andrew Greenberger was filming a new robotic competition called Depth Charge: Underwater Robot Challenge, for the Discovery Channel. Greenberger was the producer who filmed a documentary on FIRST in 1998, and then helped organize TLC's *Robotica.* His new show followed teams of students and scientists as they designed and raced underwater robots. Greenberger marveled at how far the genre had come, and recalled asking one young participant on his new program to name his hero. The student's eyes lit up as he answered, "Carlo Bertocchini." The kid had pictures of Biohazard on his wall at home.

Across the country in Florida, Trey Roski and Greg Munson were holding another student-oriented competition—the first ever "Battlebots IQ," at Universal Studios, Florida. Twenty-five teams of kids and their mentors competed in a miniature Battlebox with middleweight battle robots. The competition lasted two days, and wasn't televised. Roski wanted to build a community of high schools around his new event, then try to sell it to a major network.

Boston-area high-school teacher Mike Bastoni, once a FIRST devotee who now helped out Battlebots IQ, thought the two events could live in harmony. "When Dean said he wanted FIRST to be the Olympic Committee of Smarts, I bought it hook, line, and sinker," Bastoni said. "But there are hundreds of sports in the Olympics, and wrestling and boxing are two of them. FIRST and Battlebots complement each other. Battlebots gives us deep immersion in mechanical contests, while FIRST is about inspiring students by pairing them with engineers."

Attendees at Battlebots IQ said the spirit of camaraderie and cooperation pervaded the pits in the exact same way it does at FIRST events. They also reported the strange sight of Trey Roski dancing the electric slide with students during a break in the action.

As for the main event on the robotics sports circuit, Battlebots, the bad news spread throughout the gearhead community in early September 2002: Comedy Central wasn't going to renew its once hit show. Ratings had dropped to under 500,000 viewers per episode. Though the network would continue to air its new episodes, and then reruns, it wanted to eventually fill the time slot with something new.

Trey Roski and Greg Munson sent an e-mail to the builder community, sorrowfully explaining that plans for the annual November tournament were on hold. "Without this television funding from Comedy Central, BattleBots Inc. simply cannot afford to hold a live competition," they explained.

The cousins said that they were pursuing new avenues of funding and a possible deal with other networks. "In the end, this challenging time may turn out to be the greatest thing that ever happened to our sport," they wrote. "As we overcome obstacles and capitalize on new opportunities, BattleBots will continue to flourish for years to come. We are actively exploring some very promising options at this time. Stay tuned for an update very shortly."

The builders reacted with a mix of sadness and optimism. The community wasn't going to come together in November for the first time in four years. But they also felt there was too much enthusiasm behind the concept for it to simply disappear. "Too many people know about this and are deeply involved in it," said Stephen Felk.

In New York, Steve Plotnicki, at least, was happy. He hoped this cleared the way, finally, for *Robot Wars* to catch on in the United States. Four of the episodes filmed by Mentorn the previous year had run on Nickelodeon, and the ratings among young boys aged six to twelve were suprisingly high. "The lack of competition will allow us to try and work on building the image of the house robots as marketable characters," he said. "If the Battlebots format doesn't get on the air anywhere else, and viewers stop being confused as to whether robot combat is sportlike, then maybe we can actually get the property to be on the same track it's on in the U.K. and other countries.

"Cross your fingers for me," Plotnicki said.

Up in Novato at the Battlebots office, the mood was somber. But Roski and Munson had known for a while that the cable TV–viewing public would only watch Biohazard, Nightmare, Ziggo, and the rest of the robots play out their internecine rivalries for so long. To really make it last as a sport—to draw attention beyond the circle of dedicated gearheads and make money, they would have to magnify the spectacle of the competition.

Marc Thorpe, whose vision for robotic combat had been successful—and painful—beyond anyone's reckoning, had always talked about amplifying the power of the machines, battling three-ton robots, instead of 150-pounders. The Sci-Fi Channel once proposed a show called *Robodeath,* and even met with the WWF's Vince McMahon and Mark Pauline, but it never got off the ground.

Now Roski and Munson were going to try it. The idea was to have vaunted gearheads like Mark Setrakian, Christian Carlberg, and Carlo Bertocchini team up with such companies as John Deere and Ford, to create beastly, fire-breathing contraptions that fought in the desert. The event would be called Battlebots Extreme, and could theoretically extend the sport of robotic combat far into the future.

The cousins started planning. Someone in the Battlebots office in Novato got the idea to call Mark Pauline, to see if the SRL crew wanted to be involved. Pauline would take the call—he always listened to these offers with a bemused smile on his face. Inevitably, he would leave them hanging.

But another cycle was already beginning. They could do everything better this time around: The audience would be set farther back, away from flying debris; the rules of the contest would be so unique that they couldn't possibly be cloned by competitors; and when anyone asked that old, predictable question—whether robotic combat was violent—they could say: "Sure, it might be, but that's nothing compared to the violence of the American business and legal systems, and the human cruelty that allows people to strip each other of dignity in pursuit of greed."

Even if the new format doesn't catch on, robotic sports are here to stay. The crocodiles have escaped their cages, and no one is going to stuff them back in again.

Acknowledgments

Strangely, writing this nonfiction book required conjuring the preposterous illusion that I, too, belonged to the fraternity of the mechanically adept. For helping me weave that web, I owe an enormous debt of gratitude to Stephen Felk, Peter Abrahamson, and Scott La Valley. Each reviewed parts of this manuscript and found the legion of errors that would have destroyed my credibility with readers who know, for instance, when "a current at 24 volts" is more accurate that "a 24 volt current." The surviving mistakes are entirely my own, of course, and hint at the unfortunate truth about my gearhead status.

At least a hundred other robot builders also helped me along the way, enthusiastically sharing their stories and their outlook on the sport. There are too many names to mention, but I want to give special thanks to those who welcomed me into their homes or schools to observe their robot-building operations, or who otherwise lent invaluable support: Carlo Bertocchini, David Calkins, Patrick Campbell, Christian Carlberg, Dan Danknick, Terry Ewert and his employees at Westar Manufacturing, Nola Garcia, Ken Goldberg, Roy Hellen, Jamie Hyneman, Young Ihm, Paul Mathus, Christian Ristow, Jim Smentkowski, the Tilford family, Mike Winters, Will Wright, Mike Bastoni and the students at Plymouth North High School near Boston, and Michelle Ritter and her FIRST team at Pioneer High School in San Jose. In England, I visited with Ian Lewis, Simon Scott, and Vinny Blood of Team Razer, and Ian Watts of Team Big Brother was also extremely helpful.

Thanks also to the competition organizers who spoke with me and didn't seem to mind—too much—that I kept asking about the unpleasant conflicts of their business. Trey Roski and Greg Munson invited me to every Battlebots event in 2001 and 2002. Steve Plotnicki of Astor Holdings overcame his initial distrust of the project and was refreshingly straightforward. Dean Kamen took time from his valuable schedule for an interview, and when we didn't finish in time, allowed me to trot along and shout questions as he rode his Segway Human Transporter around the 2002 Silicon Valley FIRST Regional in San Jose. I'm still kicking myself that I didn't ask to take a test drive. Woodie Flowers, a pleasure to talk to, sent me a box full of old videotapes of MIT's famous 2.70 competition. Tom Gutteridge of Mentorn stayed late one night in his London office for a productive chat; I also want to thank Alice Triffit and her colleagues at Mentorn who facilitated my travels to the Robot Wars competitions in England.

Throughout the long process of writing this book, Marc Thorpe was overly generous with his time and assistance. From the start, he had the patience and insight to help me pursue the straightforward truth about the history of Robot Wars, rather than a favorable, sepia-toned interpretation.

At *Newsweek,* George Hackett, Mark Miller, Marcus Mabry, and Karen Breslau made sure I had the time I needed to finish the project, and my colleague Steven Levy provided valuable guidance and advice throughout the process. Photographer Gerry Gropp and photo editor Bruce Jaffe were hugely helpful in taking and collecting some of the necessary pictures. Sam Register and Jia-Rui Chong backed me up with research assistance. I'm also grateful to editor Mark Whitaker and managing editor Jon Meacham for installing me in Silicon Valley in the late nineties, and for giving me the chance to write for a great magazine.

At Simon and Schuster, Jon Malki and Jeff Neuman made the book better with their skillful editing. My agent, Frank Scatoni of Venture Literary, was the source of endless patience and sharp counsel. My friends Sean Meshorer and Hilary Black also contributed indispensable feedback.

My most heartfelt thanks are to my family, Robert, Carol, Brian and Eric; each read this book in its early stages and tolerated the bear at the other end of the phone during the months I spent grumpily writing; and, of course, to Jennifer, who matched my enthusiasm for the robots with her own, and helped me find the human drama amid all the nuts and bolts.

Photo Credits

1) Courtesy of Marc Thorpe
2) Courtesy of Marc Thorpe
3) Courtesy of SRL
4) Courtesy of SRL
5) Patrick Campbell
6) Sean Casey
7) Patrick Campbell
8) Sean Casey
9) Sean Casey
10) Sean Casey
11) Sean Casey
12) Sean Casey
13) Courtesy of *Battlebots*
14) Sean Casey
15) Sean Casey
16) Sean Casey
17) Sean Casey
18) Sean Casey
19) Sean Casey
20) Courtesy of *Battlebots*
21) Courtesy of Woodie Flowers
22) AP/Wide World
23) UPI
24) Courtesy of *Battlebots*
25) Courtesy of Mentorn
26) Courtesy of *Battlebots*
27) Courtesy of *Battlebots*
28) Sean Casey

29) Courtesy of *Battlebots*
30) Gerry Gropp
31) Jennifer Granick
32) Courtesy of Team Razer

About the Author

Brad Stone joined *Newsweek*'s writing staff in 1995 and became the magazine's Silicon Valley correspondent in 1998. Stone has covered a wide variety of stories, including the Mark McGwire–Sammy Sosa home run race, the Napster lawsuit, and the rise and fall of the dot.com economy. A graduate of Columbia University, Stone lives in San Francisco with his wife, Jennifer, their cat, and their robotic vaccuum. To learn more about this book and the wide world of robotic sports, please visit www.gearheadsthebook.com.